T0143152

EVOLUTIONARY INNOVATIONS

# EVOLUTIONARY INNOVATIONS

Edited by
Matthew H. Nitecki

*With the Editorial Assistance
of Doris V. Nitecki*

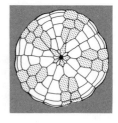

The University of Chicago Press    Chicago and London

MATTHEW H. NITECKI is curator of fossil invertebrates at the Field Museum of Natural History. He is the editor of four previous volumes based on Field Museum symposia published by the University of Chicago Press: *Evolutionary Progress, Extinctions, Coevolution,* and *Biochemical Aspects of Evolutionary Biology.*

The University of Chicago Press, Chicago 60637
The University of Chicago Press, Ltd., London
© 1990 by The University of Chicago
All rights reserved. Published 1990
Printed in the United States of America

99 98 97 96 95 94 93 92 91 90      54321

Library of Congress Cataloging-in-Publication Data

Evolutionary innovations / edited by Matthew H. Nitecki,
    with the editorial assistance of Doris V. Nitecki.
        p.   cm.
    Includes bibliographical references and index.
    ISBN 0-226-58694-4 (cloth) — ISBN
    0-226-58695-2 (pbk.)
    1. Evolution.   I. Nitecki, Matthew H.
        II. Nitecki, Doris V.
QH366.2.E869   1990
575—dc20                                        90-10990
                                                    CIP

♾ The paper used in this publication meets the minimum requirements of the American National Standard for Information Sciences—Permanence of Paper for Printed Library Materials, ANSI Z39.48-1984

# Table of Contents

# Preface

Since the time of Darwin, biologists have wrestled with the problems of evolutionary innovations. Charles Darwin, C. Lloyd Morgan, George Gaylord Simpson, Ernst Mayr, and others have all contributed to the debate, while this volume is a cross section of present-day ideas. However, it is certainly not a synthesis – such a synthesis is not yet possible – but rather a survey of biological ideas from Darwin to now, and an introduction to the theoretical and conceptual aspects of the contemporary state of knowledge on evolutionary innovations.

This book is based on the Eleventh Annual Spring Systematics Symposium, *Evolutionary Innovations: Patterns and Processes,* held at the Field Museum of Natural History in Chicago on May 14, 1988. Sincere thanks go to those individuals whose cooperation and knowledge made the symposium a success, and to those who helped with the book. Joel Cracraft must be singled out for his advice on matters evolutionary, help with the choice of speakers, and for his patience with our ignorance. National Science Foundation defrayed partial costs of the symposium through grant BSR-8823014.

We are greatly indebted to many persons for their wise counsel and aid with the manuscripts: Stevan J. Arnold, George A. Bartholomew, John R. Bolt, William C. Burger, John J. Flynn, Jack Fooden, Lance Grande, Raymond B. Huey, David Jablonski, Russell Lande, Scott Lidgard, Eric R. Lombard, Malka Moscona, Jack H. Prost, J. John Sepkoski, Jr., Janice B. Spofford, William D. Turnbull, James W. Valentine, Leigh M. Van Valen, Michael J. Wade, and William Wimsatt.

The Symposium Coordinators were Kristine Bradof, who saw that everything worked smoothly during the symposium and patiently worked on the composition of the book, and Sophia Brown, who cheerfully tended to the final product. Zbigniew Jastrzebski helped with figures.

# Contributors

William Bemis, Zoology Department, 348 Morrill Science Center, University of Massachusetts, Amherst, MA 01003

David J. Bottjer, Department of Geological Sciences, University of Southern California, Los Angeles, CA 90080-0741

Warren Burggren, Zoology Department, 348 Morrill Science Center, University of Massachusetts, Amherst, MA 01003

Brian Charlesworth, Department of Ecology and Evolution, 1101 East 57th Street, The University of Chicago, Chicago, IL 60637

James M. Cheverud, Department of Anthropology, Northwestern University, Evanston, IL 60208. Present address: Department of Anatomy and Neural Biology, Washington University School of Medicine, St. Louis, MO 63110

Joel Cracraft, Department of Anatomy and Cell Biology, Post Office Box 6998, University of Illinois, Chicago, IL 60680, and Department of Zoology, Field Museum of Natural History, Chicago, IL 60605

David Jablonski, Department of Geophysical Sciences, The University of Chicago, 5734 S. Ellis Ave., Chicago, IL 60637

Jeffrey S. Jensen, Museum of Comparative Zoology, Harvard University, Cambridge, MA 02138

Karel F. Liem, Museum of Comparative Zoology, Harvard University, Cambridge, MA 02138

Gerd B. Müller, Department of Anatomy, University of Vienna, Währingerstrasse 13, A-1090 Vienna, Austria

Matthew H. Nitecki, Geology Department, Field Museum of Natural History, Roosevelt Road at Lake Shore Drive, Chicago, IL 60605

Annette L. Parks, Institute for Molecular and Cellular Biology and Department of Biology, Indiana University, Bloomington, IN 47405

Brian A. Parr, Institute for Molecular and Cellular Biology and Department of Biology, Indiana University, Bloomington, IN 47405

Rudolf A. Raff, Institute for Molecular and Cellular Biology and Department of Biology, Indiana University, Bloomington, IN 47405

Adolf Seilacher, Universitat Tübingen, Institut und Museum für Geologie und Paläontologie, Sigwartstrasse 10, 7400 Tübingen 1, West Germany, and Kline Geology Laboratory, Yale University, P.O. Box 6666, New Haven, CT 06511

Gregory A. Wray, Institute for Molecular and Cellular Biology and Department of Biology, Indiana University, Bloomington, IN 47405

# INTRODUCTION

# The Plurality of Evolutionary Innovations

*Matthew H. Nitecki*

Innovations cannot be explained by any single cause. There is much new evidence and many new ways of conceptualizing the problem of evolutionary innovations, and research on this subject is accordingly very heterogeneous. Therefore, and for the simple sake of convenience, we divide our study of evolutionary innovations into five different but not unrelated approaches: (a) ontological, (b) genetic and developmental, (c) morphological and physiological, (d) paleontological, and (e) hierarchical causes and effects of innovations, none of which is sufficient in itself. Though their emphases may differ, these approaches are complementary. However, even taken together they may not fully satisfy our need to explain.

Clearly we must all speak the same language when we describe or define evolutionary innovations. Only then can we postulate a general theory or construct a model to explain the totality of evolutionary innovations. But this is no easy task. Almost from the beginning we encounter difficulties. While model-making may be quantitative and predictive, it still wears a fashionable statistician's label, perhaps because it presupposes that complex problems can be reduced to simple definitions and rules. But life is more complex than any model. The simplest model usually applied to life is that of a machine. New machine models are blossoming in computer simulations and with applications of the second law of thermodynamics to evolution. But life, in a constant state of change, is a more fortuitous concourse of processes then any such simple analogy. It is a *process*, not an entity, occurring at many levels from the chemical to the social organizations of human societies. It occurs on many spatial and temporal scales. Models of life, however sophisticated, cannot now, nor perhaps ever, represent all the tremendous complexities of life.

Although each contributor to this volume attempts to explain the meaning of evolutionary innovations, we are not able to define the concept precisely. It is used simply to describe observed, or assumed, structures and phenomena characteristic of new biological processes and patterns. We must clarify the terminology applied to the idea of evolutionary innovations and make clear that we are concerned with biological novelties such as new taxa, and the processes and patterns responsible for their origin. Unfortunately, even to the contributors to the present volume, these terms (such as innovation, novelty, or emergence) have various meanings.

In the last few years innovation has been refined to include both processes and the resulting structures. While innovations still mean what they meant to Mayr (1960), they perhaps can be rephrased as "heritable novelties which spring from phylogenetic and developmental constraints and may serve as vehicles for carrying taxa across valleys in adaptive landscapes" (Patterson 1988:86). Thus they are the engines of evolution, its work horses. Evolutionary innovations are rarely sharply delimited, and hence their definitions are often confusing. McKinney (1988) pointed out that interpretations of evolutionary innovations depend mostly on the perspectives and perceptions of observers, and that, therefore, evolutionary innovations are concepts used differently by different people.

It is clear that the term innovation has many facets. Is it the appearance of *new* whether it be structures, processes or behavior? What is meant by "new"? Is it more than reshuffling and rearranging? Is it predictable? How do changes occur? Are they the result of many small steps, that is, continuous, slow, and gradual transitions? Or do they occur in a large, single, sudden, and unexpected step? Are some more probable, and others improbable? Is the mechanism of evolutionary innovation solely random genetic fluctuation, or are other processes at work? In his chapter on the evolution of flowers and butterflies, Charlesworth discusses arguments in support of small steps that originate innovations. Raff et al. consider the question "What causes evolutionary changes in developmental processes?" Müller believes that an internal mechanism is responsible for at least some innovations.

Evolutionary innovations, in addition to the origins of morphological novelties, may include new functions, and, therefore, the physiology, of new properties. In short, evolutionary innovations involve processes and products, and form and behavior.

It is commonly assumed that modern biology in general emphasizes processes and functions over form and morphology. But the story of evolution is read from morphological changes. How organisms function, behave, or evolve new structures is dependent on morphology. Breathing, reproducing, swimming, flying, burrowing, building nests, are all "behavioral" properties with unique characteristics – the results of evolutionary innovations – entirely controlled by lungs, gonads, fins, wings, and so on. The anatomical entities, from the structure of the DNA molecule on up, determine biochemistry, physiology, and behavior.

### History of Study

The interest of biologists in evolutionary innovations is not new, but thoughtful and sensitive work on this subject (if not necessarily under this exact title) goes back well into the nineteenth century, particularly to Charles Darwin, and has been the major preoccupation of evolutionary biologists ever since.

**Charles Darwin.** Many subdisciplines of biology included innovations within their subject matter. Innovations were among the main concerns of Darwin. The conventional wisdom from Darwin's time holds that most innovations are good in that they are a necessary pool from which natural selection chooses the most beneficial. This view is still valid today. However, evolutionary innovations themselves are subject to evolutionary changes.

Ernst Mayr (1960) rekindled interest in innovations. It was he who pointed out that Charles Darwin in his sixth edition of *The Origin of Species* was concerned with innovations. Darwin asked, "Is it possible that an animal having, for instance, the structure and habit of a bat, could have been formed by the modification of some other animal with widely-different habits and structures?" (1883:133). Darwin, as usual, provided numerous examples (1883:139ff) of flying squirrels, flying lemurs, birds, and many other animals that have been replaced by "more perfect" forms which evolved to newer habitats. Structures change to suit new habits, and habits change to suit new structures, but which occurs first? Darwin claimed that the changes are simultaneous.

Darwin (1883:143-46) claimed that "organs of extreme perfection" are formed by continuous perfection through gradation. Modes of transition are "Organs ... formed by numerous, successive, slight modifications" (1883:146). He discusses numerous difficulties encountered with innovations. "*Natura non facit saltum* ... or as Milne Edwards has well expressed it, Nature is prodigal in variety, but niggard in innovation. Why ... should there be so much variety and so little real novelty?" (1883:156).

Darwin's greatest contribution to the understanding of innovations has been the ability to explain that adaptations originate by natural selection of innovations. Furthermore, he assumed that these innovations vary, are heritable, and are selected for.

**Post-Darwin.** I disagree with Mayr (1960:350) that in the first half of the twentieth century the question "How does an evolutionary novelty emerge?" was not asked. The discussions regarding emergent evolution were concerned with the origin of innovations, and whether the innovations are deterministic. Weismann (1904:346) thought they are necessary!

It was Romanes (1897) who, almost a hundred years ago, after the fury of earlier Darwinian controversies was dying out, argued that the theory of natural selection, mainly concerned with adaptation, and evolutionary innovations are the kernel of evolution. Romanes's arguments are worth recalling. He claimed that only a few systematists held that the chief merit of Darwin's theory was the explanation of the origin of species. The idea of the origin of species is "but a survival, or a vestige, of an archaic system of thought" (1897:159). Only when species were believed to be the result of separate acts of creation was Darwin's theory explaining their origin by means of natural selection of such paramount importance. Indeed, this was the reason for the *Origin*'s success. But Romanes argued that since the *facts* of evolution were proven [*sic*], it is the *method* that must be explained. However, natural selection covers a much larger area than adaptation. Species by themselves lost their deep philosophical significance. Of real importance is that evolution is "a theory of the origin and cumulative development of adaptations – whether structural or instinctive, and whether the adaptations are severally characteristic of species only, or of any of the higher taxonomic divisions" (1897:161). Thus the theory of evolution is a theory of adaptations applied to the entire nature irrespective of hierarchical level. Natural selection is exclusively responsible for adaptive characters. The adaptive characters of Romanes are our evolutionary innovations. That is why Romanes claimed that the theory of evolution is mainly the explanation of the origin (cumulative development), existence, and spread (diffusion) of evolutionary novelties, and only secondarily of the origin of taxa. Romanes devoted half of his book to how novelties originate. Like almost everyone who has dealt with innovations, Romanes started his discussion with the innovation of the bird's wing. Unfortunately, many of his later arguments lose much of their impact because of his support of modified Lamarckism.

Weismann (1917) was so preoccupied with innovations as to claim that everything depends upon them and that whole orders of animals are, so to speak, *made up of innovations*.

**George Gaylord Simpson.** Simpson's *The Major Features of Evolution* (1953) forms the principal foundation upon which the modern concept of evolutionary innovations is based, and from which it makes its beginning. Simpson's arguments of how *key mutations* cause a lineage to enter a new *adaptive zone* and subsequently to undergo an adaptive radiation there seems to be the direct precursor to the concept of innovation criticized by Cracraft (this volume) and elaborated as *symecomorphosis* by Liem (this volume).

**Ernst Mayr.** The study of evolutionary innovations as a subset of evolutionary biology is very recent. It began with Simpson but was fully developed in Ernst Mayr's (1954) key publication on evolutionary novelties and continued with his 1960 paper on the emergence of evolutionary innovations. It is Mayr, more than any other living biologist, who has established evolutionary innovations as a distinct subject matter in its own right.

Indeed, Mayr is responsible for the great progress in the study of evolutionary innovations, and has pushed these studies into the successful area of advance work in the theoretical aspects of evolutionary biology. As is clear from the chapter by Burggren and Bemis, some methodologies may be new in their field of specialty, physiology, while others come from sister disciplines.

Of great interest, however, is that much of the work in evolutionary biology still follows the "modern synthesis" path of close association among systematists, population biologists, geneticists, paleontologists, physiologists, embryologists and others. The present volume demonstrates that many of these disciplines continue to contribute to the theoretical aspects of evolutionary biology.

Mayr includes in evolutionary novelties "any newly arisen character, structural or otherwise, that differs more than quantitatively from the character that gave rise to it," or more briefly "any newly acquired structure or property which permits the assumption of a new function" (1960:351). Mayr, in 1963, is again concerned with "inventive" evolutionary novelties (1963:602-

21) and again gives as examples feathered wings, the middle ear, and swim bladders among others. He defines an evolutionary novelty as "any newly acquired structure or property that permits the performance of a new function, which in turn, will open a new adaptive zone" (1963:602).

Mayr sees three modes of origin of novelties (1963:602-3): (1) phenotypic by-product of a genotype selected for other reasons; (2) Severtzoff's "intensification of function," which is simply a modification of preexisting structure; and (3) "change of function," which he attributes to Darwin. Such a structure must have the capacity to perform two functions and be able to duplicate functions. Furthermore, he claims that "a shift into a new niche or adaptive zone is, almost without exception, initiated by a change in behavior . . . the [structural] adaptations . . . are acquired secondarily" (1963:604).

## Ontology: Nature of Evolutionary Innovations

While all of us are interested in the origin of innovations, Joel Cracraft, in his most conceptual chapter, is particularly concerned with this question. He delineates the pattern and history of different evolutionary units, defines sample evolutionary innovations, and erects a phylogenetic hypothesis to explain his sample. To achieve the highest degree of accuracy, he chooses groups with minimal extinction and determines how mutational events affect the phenotype. Cracraft wants to identify these innovations and to establish the hierarchical level at which evolutionary taxa are classified, at which level they permit quantification of rates of change along different lineages, and thus permit the identification of historical patterns of character covariance. He asks whether changes in diversity are caused by changes in the rates of speciation and extinction. Are these changes caused by increases in the rate at which populations become isolated, in the rate at which novelties enter the population, or in the rate at which these novelties become fixed? Cracraft claims that patterns are produced at levels of individual ontogenies and of differentiating populations, and, therefore, the dynamics of patterns are at the molecular/developmental and population levels. Patterns at higher levels are probably epiphenomena of extinctions and sampling errors.

Cracraft's chapter is in almost direct opposition to arguments made by Liem on the role of key innovations, and by Jablonski and Bottjer on the relationships between evolutionary phenomena observed at different taxonomic levels. He goes a long way toward making more precise the systematic methodology used to study innovations.

### Genetics and Development

**Population Biology.** Brian Charlesworth embarks upon the arguments on the genetic basis of evolutionary change versus the nature of morphological evolutionary transformation. He asks basic questions about the broadest aspects of the three alternative models of new adaptations. To Charlesworth the evolutionary genetics of adaptation means that in order for any system to work well as a whole, all its components must be integrated. Charlesworth asks how new complex adaptations arise. Do they arise according to (1) Fisher's model in a series of steps each more advantageous than the previous one; (2) the Goldschmidtian model in which mutations produce fully functional complexes at once; or (3) the Sewall Wright model of shifting balance, which involves an interaction between the effects of drift and selection to cause a population to evolve toward a state of higher fitness that would not be accessible through the effects of drift or selection alone? Charlesworth argues that the Darwinian model best explains the origin of evolutionary innovations.

**Embryology.** Rudolf Raff and his coworkers are concerned with interesting new perspectives on the mechanisms of evolutionary changes in development, particularly early development and heterochrony in sea urchins. The main focus of their chapter is on developmental mechanisms. They argue that morphological changes are possible only when the ontogeny and the genes responsible for ontogeny are modified. In a series of imaginative investigations they compare theoretical predictions about the nature and mechanisms of developments with experiments on developmental mechanisms. They set strict guidelines for their investigations, and demonstrate how the evolution of embryonic cell lineages develops. They find great flexibility in early development, a mosaic in the evolution of cell lineages which demonstrates well-defined heterochronies, and the presence of nonheterochronous mechanisms as well.

Gerd Müller complements Charlesworth's morphological and developmental issues and their relationship to genetic changes. He is also interested in developmental mechanisms and plasticity, and approaches the origin of morphological innovations through the study of the mechanisms of development. Müller presents the distinction between genetic and morphological evolution. He argues that morphological innovations and the consequent appearance of morphological structures are controlled by the limited number of mechanisms and the dynamics of development. To test the theory with experiments, he uses the development of the vertebrate limb. Müller demonstrates that small changes in the processes of development may cause major morphological changes. Of particular interest to him is the role of the intrinsic organismic factors on the pattern and direction of evolutionary changes. However, not all innovations need be adaptively explained; instead they can be explained by historically acquired developmental rules that control the lineage-specific range of possible structural change. He is very successful at integrating the issues of heterochrony, raised by Raff and his coworkers, and at integrating heterochrony with the general theme of evolutionary innovations.

## Morphology and Physiology

It is very important to point out the potential roles of genetic and developmental *processes* in introducing novelties that are seen by morphologists and physiologists as *pattern*. On the other hand these processes appear to paleontologists as cumulative or historically recognized patterns.

**Quantitative Morphology.** James Cheverud, in his rather formalistic chapter about hereditary constraint on evolutionary innovations, is interested in the evolution of morphological innovation through morphological integration. He views genetic structures (and their interaction with stabilizing selection) as imposing constraints on the evolution of development and morphology, thereby producing stable, long-term patterns of association among characters. Cheverud uses studies of macaques and baboons to confirm his theoretical considerations of levels of evolution of morphological integration patterns. He does this by investigating the pattern of phenotypic correlation and by using element scaling analyses. His studies clearly show that patterns of phenotypic correlation and the size of phenotypic variance are constant over millions of years. Thus morphological integration in these organisms is very stable, and innovations do not originate through changes in the patterns of morphological integrations. Cheverud's rigorous mathematical deductions are similar to the intuitive statements of others in this volume; however, he is the only author to develop the quantitative genetic approach to the subject of constraint.

**Adaptive Morphology.** Karel Liem's chapter is the most central to the major issue of the book. It is an extension and elaboration of the evolutionary model that he developed in earlier papers, invoking the role of key innovative structures in morphological evolution and diversification. Liem makes an impressive use of the concepts of ecological and genealogical hierarchies. He recognizes that most novel structures are generally inconsequential. The key evolutionary innovations are those that lead to major evolutionary shifts, and are initially new structures that eventually lead to new functions. These key innovations allow the invasion of new and major habitats, and are associated with adaptive radiations. To Liem *natura facit saltus* (in a current sense of *saltus*) following the emergence of key innovations. He correlates big environmental changes with big evolutionary changes. Liem's knowledge of the functional morphology of fishes allows him to support his ideas with examples from labroid fishes. Structure and function must be well integrated with ecological factors for the key innovations to initiate evolutionary diversification. Liem calls this matching of morphological, functional, and ecological factors *symecomorphosis*.

Jeffrey Jensen tests historical hypotheses and further develops and tests Liem's ideas of evolutionary innovations and key innovations by using sister-group analysis. His chapter also relates very closely to the issues raised by Cracraft. Comparing a taxon that possesses an evolutionary innovation with a sister group lacking the innovation avoids the pitfalls of ad hoc explanations. Jensen also uses examples from fishes to provide a tool for analyzing the influence of intrinsic features of design on evolution.

**Physiology.** Warren Burggren and William Bemis study the paradigms and pitfalls of physiological evolution and how comparative physiologists study evolution, and call for ambitious changes in the paradigms guiding physiological research. They accuse physiologists of not integrating their data with the mainstream of modern evolutionary biology, by not creating good evolutionary scenaria for their data. Of the many reasons for that situation, some are historical or traditional while others relate to interpretation of the structure-function relationship. Burggren and Bemis see great differences between morphology and physiology, mainly because minute differences in anatomy yield great differences in physiology. Physiological processes are only weakly linked to observed structures, and evolutionary physiologists have no benefit of the fossil record.

Burggren and Bemis argue for the emerging field of evolutionary physiology. They examine how evolutionary biologists study evolution, how evolutionary morphologists study functional characters, and what ideas to apply to
the study of physiological evolution.


## Paleontology


The fossil record underlies all historical biology. Almost any random examination of the fossil record reveals endless examples of the appearance of
new characters. Prior to the twentieth century, behavior was mostly considered not preservable in the fossil record. In this respect paleontologists
concerned with innovations differed from other biologists. Most modern
paleontological ideas about innovations are traceable to George Simpson,
who demonstrated the conceptual agreement between the patterns of the
biological present and the paleontological past. A series of elegant new
studies based on theoretical models and clearly defined theoretical problems
are now regarded as appropriate subjects of experimental and observational
analysis. Such studies are represented by the theoretical problems and
models in chapters by Seilacher and by Jablonski and Bottjer.

**Constructional Morphology.** Adolf Seilacher has changed our ideas about
the fossil record by showing that the behavior of extinct organisms can be
read from the rocks. To Seilacher evolutionary adaptation, within the constraints of phylogeny and morphology, utilizes and optimizes innovation
potentials present in all biological structures. In his chapter, he demonstrates
that evolutionary adaptations may be opportunistic rather than elegant,
planned solutions. The sand-dollar syndrome, a set of multiple evolutionary
innovations, features a bent-over-backwards switchover for what was the
historic mistake (in hindsight) of locating the mouth under the body. The
change from bottom feeding to gravitational sieving, which required channel-

ling food particles from the upper surface to the lower, was achieved by lunules and "banana-road" systems of food grooves that developed independently at least four times. Seilacher demonstrates the inevitability of innovation, particularly when a structure reaches a new ecologic niche, the importance of alternative morphogenetic pathways, and the significance of paleogeographic data for cladistic reconstructions.

**Fossil Record.** David Jablonski and David Bottjer explore the ecology of evolutionary innovations in the fossil record, where there is abundant data on the timing, sequence, geography, and ecology of innovations. In a general sense their paleontological data seeks evolutionary driving forces. They study the rates of variations and origination as a function of niches. In a very imaginative and thought-provoking way they characterize the broad, non-specific attributes of evolutionary innovations by counting densities of first occurrences in the fossil record.

In the past the studies of innovation have focused on two elements: the ecological characteristics that might initiate novelty and the environments that might foster it. Jablonski and his colleagues (e.g., Jablonski et al. 1983; Jablonski and Bottjer 1983; see also Sepkoski and Sheehan 1983; Sepkoski and Miller 1985) have found some interesting patterns. For benthic marine organisms, for example, onshore habitats appear to be the source of evolutionary innovation, and offshore habitats appear to contain archaic taxa and modes of life. It is also known that many taxa restricted to outer-shelf, slope, and deep-water habitats originated in, or previously occupied, shallow waters.

More higher taxa of post-Paleozoic invertebrates originated in shallow-water environments than would be expected by chance. However, at the genus and family levels within those higher taxa, there is no environmental bias in origination, apparently in accordance with the standing diversity gradients of their groups. The higher-taxon pattern is not an artifact of preservation or sampling, but rather indicates that different processes of origination control different hierarchical levels. The contrast between patterns of innovation observed at different taxonomic levels is in direct contrast to arguments presented by Cracraft. Both opposing viewpoints are appropriate, especially since Jablonski and Bottjer present their arguments effectively with the empirical basis.


## Conclusions


We are right in rejecting the notion that innovations are always creative and progressive. But are we also right to reject (consciously or otherwise) that all history, evolutionary or otherwise, is progressive? Can innovations not be equated with evolutionary progress itself? Nineteenth century progressive veneers can be rejected, but surely there is more to progress than that.

What is seen in the fossil record (or *interpreted* as seen in the fossil record) is an abundance of such radical innovations as the appearance of new taxa, new structures, or invasions of new environments. It seems as if the existence of innovations in the fossil record is indisputable. Furthermore, in order to prove that innovations are not chance events, it would be necessary for us to predict Silurian innovations, for example, from our knowledge of Ordovician paleontology. In other words, we should be able to predict Silurian life from the Ordovician perspective. However, no one has demonstrated that such predictions are possible, so it is easy to assume that the appearance of innovations has been controlled by chance.

There are two main themes in our book: adaptations and phylogeny. Phylogenetic control seems to be a common theme uniting our discussion of innovations. No innovation can be discussed without considering history. However, the time scales of paleontologist and neontologist appear to be different. But most of the differences between innovations and *key* innovations may be semantic. We must always remember that when talking about innovations, novelties, and emergences, we are really talking about ideas. We sometimes assume that we understand the underlying processes, which we name. But in effect we are describing our interpretations: the more novel the interpretations, the more novel the ideas and processes appear.

The basic ideas of innovation are still being debated. To establish innovation on more solid foundations, more study is needed and many questions must be answered. Conceptions of evolution may have changed since Darwin, but the basic ideas of Darwin remain unchanged. We now ask more detailed questions about innovations. However, we are still very far from the synthesis or total integration of our observations.

## References

Darwin, C. 1883. *On the Origin of Species*. New ed. from 6th ed. New York: D. Appleton and Co.

Jablonski, D., J. J. Sepkoski, Jr., D. J. Bottjer, and P. M. Sheehan. 1983. Onshore-offshore patterns in the evolution of Phanerozoic shelf communities. *Science* 222: 1123-25.

Jablonski, D., and D. J. Bottjer. 1983. Soft-bottom epifaunal suspension-feeding communities in the Late Cretaceous: Implications for the evolution of benthic paleocommunities, pp. 747-82. In *Biotic Interactions in Recent and Fossil Benthic Communities,* ed. M. J. S. Tevesz and P. L. McCall. New York: Plenum.

Mayr, E. 1954. Change of genetic environment and evolution, pp. 157-80. In *Evolution As A Process,* ed. J. Huxley, A. C. Hardy and E. B. Ford. London: George Allen & Unwin Ltd.

Mayr, E. 1960. The emergence of evolutionary novelties, pp. 349-80. In *Evolution After Darwin,* ed. S. Tax. *The Evolution of Life.* Vol. 1. Chicago: The University of Chicago Press.

Mayr, E. 1963. *Animal Species and Evolution.* Cambridge: Harvard University Press.

McKinney, F. K. 1988. Multidisciplinary perspectives on evolutionary innovations. *Trends in Ecology and Evolution* 3: 220-22.

Patterson, B. D. 1988. Evolutionary innovations: Patterns and processes. *Evolutionary Trends in Plants* 2: 86-87.

Romanes, G. J. 1897. *Darwin, and After Darwin. II. Post-Darwinian Questions. Heredity and Utility.* 2d ed. Chicago: The Open Court Publishing Co.

Sepkoski, J. J., Jr., and A. I. Miller. 1985. Evolutionary faunas and the distribution of Paleozoic marine communities in space and time, pp. 153-90. In *Phanerozoic Diversity Patterns,* ed. J. W. Valentine. Princeton: Princeton University Press.

18     *Matthew H. Nitecki*

Sepkoski, J. J., Jr., and P. M. Sheehan. 1983. Diversification, faunal change, and community replacement during the Ordovician radiations, pp. 673-717. In *Biotic Interactions in Recent and Fossil Benthic Communities,* ed. M. J. S. Tevesz and P. L. McCall. New York: Plenum.

Simpson, G. G. 1953. *The Major Features of Evolution.* New York: Columbia University Press.

Weismann, A. 1904. *The Evolution Theory.* Vol. II. Translated by J. A. Thomson and M. R. Thomson. London: Edward Arnold.

Weismann, A. 1917. The selection theory, pp. 23-86. In *Evolution in Modern Thought,* By Haeckel, Thomson, Weismann, and Others. New York: Boni and Liveright, Inc.

# NATURE OF EVOLUTIONARY

# INNOVATIONS

# The Origin of Evolutionary Novelties:
# Pattern and Process
# at Different Hierarchical Levels

*Joel Cracraft*

The study of innovation occupies a central position within evolutionary biology. The concept of evolutionary innovation, or novelty, has been interpreted in many different ways by biologists, but it is probably accurate to say that most understand innovations to mean the appearance of new characters or structural/functional complexes. These innovations are found to characterize individual organisms within a population, basal species-level taxa, or they are shared by clusters of species, which may or may not comprise a monophyletic group. Generally, however, biologists have discussed the evolutionary significance of innovation within the context of large-scale manifestations of the phenotype, particularly those morphological, behavioral, or physiological traits that are interpreted to facilitate a new way of making a living and that somehow are thought to be causally related to the success of the major taxonomic group possessing them. Thus, Mayr (1963:602) has defined evolutionary novelty "as any newly acquired structure or property that permits the performance of a new function, which, in turn, will open a new adaptive zone." A similar conception of evolutionary innovation is echoed by many within the evolutionary literature. The list of identified novelties is endless, virtually equaling the diagnostic characters of all successful higher taxa. Some of those most often mentioned as "classic" examples of evolutionary innovation include: the bony skeleton of vertebrates, jaws of gnathostomes, the amniote egg, avian flight, feathers, continuously growing incisors of rodents, large brain of hominoids, the artiodactyl tarsus, the insect wing, rigid skeletons and complex spicules of sponges, and the insect pollination system of angiosperms.

The study of innovations has occupied center stage within evolutionary biology for a number of reasons. Perhaps the most important has been the desire of biologists to explain the apparent design of organisms by relating the structural/functional characteristics of these innovations to particular ways of life using the conceptual framework of adaptationism. Such an exercise, it is often suggested, informs us as to how and why the particular innovation originated and was increasingly perfected as adaptive design. That this conceptual framework has dominated the study of innovations is easily demonstrated by recasting it in terms of its converse and compiling a list of all those innovations that are considered to be less efficient or less well designed functionally relative to the more primitive condition. Whether or not such innovations exist, biologists have seldom conceptualized the problem in this way. As a result, that list would be short indeed (even reductions or vestigial structures are regularly given an adaptive interpretation).

A second major reason why the study of innovation has been considered so important is the conjecture that innovations are causally related to the evolutionary success – in terms of diversity or historical persistence – of clades. The concept of "key innovations" continues to play a prominent role in explanations of this type, and all the examples just given have been identified, by one or more authors, as key innovations.

This chapter will evaluate the concept of key innovation and assess the proposition that these features are causally important in explaining patterns of diversification. The discussion of these subjects will be organized around three main themes, one ontological, one methodological, and the third empirical:

1. *The ontological theme.* A substantial proportion of the key innovations noted in the evolutionary literature are typological constructs. Empirically they do not represent true evolutionary innovations, and as such are

limited in what they can tell us about the processes actually responsible for the origin and maintenance of evolutionary novelties. The ontological mistake has arisen because the analysis of evolutionary innovation has largely been confined to hierarchical levels well above that at which processes producing innovations have operated.

2. *The methodological theme.* A more rigorous methodological approach to the study of innovation is required, one that begins at the hierarchical level of speciation analysis and extends downward to molecular developmental genetics. Manifestations of pattern above the level of differentiating populations, or basal taxa, are epiphenomena, or effects, of lower level processes.

3. *The empirical theme.* A conjecture will be proposed: the role ascribed to key innovations as being causally related to patterns of diversification must be largely incorrect. This is not to say that innovation, analyzed at its proper hierarchical level, has no influence on the subsequent history of a group, but that the causal connections are complex. Partitioning different patterns and processes into their proper hierarchical levels will help resolve these complex causal connections.

### The Historical Anatomy of a Key Innovation: Avian Flight

Structural complexes are typically identified as key innovations because, first, they are characteristic of, or correlated with, a monophyletic taxon of high diversity, and second, they are judged to be functionally important in the way of life of species in that taxon. Most important, perhaps, is the supposition that the innovation itself caused the high diversity. The origin of avian flight is frequently identified as one such innovation (Mayr 1963:598; Futuyma 1986:356) and can serve as an exemplar of most kinds of key innovations discussed in the literature. One approach to understanding the possible role key innovations may have played in the success of a group is to examine the

*historical structure* of the putative innovation within the context of a phylogenetic hypothesis for the group and its close relatives (see also Lauder 1981;
Larson et al. 1981; Stiassny and Jensen 1987; Lauder and Liem, in press).

The cladistic pattern of early avian relationships is becoming better
understood and can be used to reconstruct the historical structure of avian
flight. The phylogenetic hypothesis of figure 1 is based on numerous characters, including some not directly associated with flight (Cracraft 1986, 1988),
but only those characters comprising the flight apparatus will be considered
here. Without belaboring the point, categorizing structures as part of something called the "flight apparatus" is often arbitrary. Many modifications of
the vertebral column, pelvis, and hindlimb undoubtedly have been important
functional components of avian flight since their inception, yet they will be
omitted from this discussion.

The relationships between birds and one or more groups of theropod
dinosaurs is now well established (Ostrom 1976; Padian 1982; Gauthier and
Padian 1985; Gauthier 1986), and present evidence suggests that the sister
group of birds is the Deinonychosauria (Gauthier 1986). Within the context
of this phylogenetic hypothesis, then, it is apparent that some characters of
the flight apparatus arose prior to node 1 of figure 1, including elongated
forelimbs, a thin and bowed metacarpal III, and, most probably, an ossified
sternum (Gauthier 1986). Features postulated to have arisen at the hierarchical level of node 1 include feathers, loss of fusion between the scapula
and coracoid, rotation of the forelimb so that the elements can assume a typical avian resting position (Martin 1985), and an enlarged forebrain, midbrain, and cerebellum. *Archaeopteryx*, it should be pointed out, was almost
certainly capable of true powered flight.

Avian remains recently discovered in the Lower Cretaceous of Spain
(Sanz et al. 1988) seem to represent a taxon that is the sister group of all
birds excluding *Archaeopteryx*. That clade (fig. 1, node 2) is defined by four
characters associated with the flight apparatus: pygostyle, strutlike coracoid
articulating with the sternum and forming a brace for the pectoral girdle and

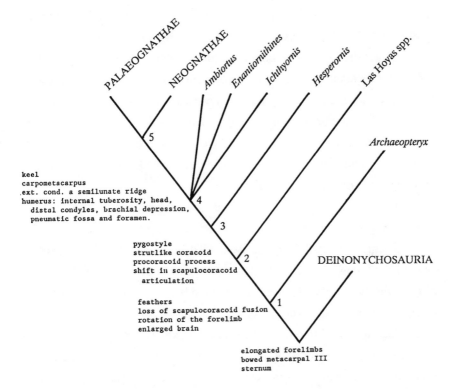

Figure 1. A phylogenetic hypothesis for the early lineages of birds showing the distribution of "innovations" associated with the "flight apparatus" (see Cracraft 1986 for more details). See text for discussion.

forelimb, a procoracoid process on the coracoid, and a shift in the scapulo-coracoid articulation to a position below the distal end of the coracoid.

At the level of *Hesperornis* and the Carinatae (fig. 1, node 3), no characters involved in flight are known to have arisen. *Hesperornis* was secondarily flightless, and the forelimb was structurally simplified. In contrast, the Carinatae (fig. 1, node 4) are characterized by a number of features that are components of the flight apparatus: a keel on the sternum, a carpometacarpus, the external condyle of the ulna developed as a semilunate ridge, and a humerus with a strongly developed internal tuberosity, strongly defined head and distal condyles, brachial depression, and a pneumatic fossa and foramen.

No new features of the flight apparatus have been shown to characterize the Neornithes or modern birds (fig. 1, node 5), though some undoubtedly exist.

Traditional descriptions of avian flight as an evolutionary innovation begin with *Archaeopteryx* and a modern flying form and create a scenario of how the structural features of *Archaeopteryx* might have been transformed by natural selection into those of a modern flying bird. This is not meant to imply that these workers envisioned all characters to be changing uniformly through time, but without a cladistic hypothesis for all relevant taxa the historical pattern of character innovation cannot be reconstructed.

Even though the majority of discussions of key innovations have been undertaken in the absence of a cladistic analysis, there are other more serious problems with traditional analyses. First, it is readily apparent that the key innovation just discussed lacks any ontological status as an evolutionarily discrete novelty. Avian flight, even if it could be said to characterize modern birds, is composed of a very large suite of characters that arose within the avian clade at very different times, extending over perhaps 50 million years or more. In addition, many of the "characters" just discussed are undoubtedly themselves aggregates of characters, which arose at different times (e.g., "feathers"). Most, if not all, of the "key innovations" listed earlier lack ontological status as discrete evolutionary novelties.

It is also important to realize that hypotheses about the sequence of origination for these characters are dependent upon the taxic sample and the phylogenetic information specified by all other characters in the analysis. In an analysis of avian flight, for example, a large number of cranial and postcranial characters contribute to establishing the phylogenetic hypothesis and thus have a marked influence on how the characters of the flight apparatus are interpreted historically.

An example such as this raises two additional difficulties for the analysis of key innovations. The first is that extinction has been rampant and it has torn apart the phylogenetic nexus, leaving a seriously inadequate sample of taxa and their accompanying characters; this sample constitutes the only evidence we have to reconstruct the history of any key innovation. This difficulty is magnified by another, namely, that a causal analysis of the evolution of key innovations must be at the level of species or below (e.g., Larson et al. 1981). Processes contributing to the evolution of novelties do not have direct causal effects at levels above the species, and thus most analyses of key innovations, which investigate patterns among supraspecific taxonomic units, are descriptively inadequate purely as a consequence of extremely poor sampling of species-level taxa in the historical record. Nearly all of the "key innovations" noted above would undoubtedly fall into this category.

To emphasize these latter points, it is worthwhile reviewing the mechanisms by which evolutionary innovations arise and become established.

### The Origin of Innovations: Intersection of Two Hierarchical Levels

The preceding discussion made the ontological claim that most so-called "key innovations" are not true evolutionary innovations but are instead aggregations of these innovations, which arose sequentially over vast stretches of time. In order to establish a more precise ontology for investigating questions about the causes responsible for the origin and subsequent history of innovations, we need to define some terms explicitly.

1. *Prospective innovations* are singular phenotypic changes that arise in individual organisms within a population as a result of a modification in one or more ontogenetic pathways.

2. *Evolutionary innovations* are singular phenotypic changes that, subsequent to arising in individual organisms, spread through a population and become fixed, thus characterizing that population as a new differentiated evolutionary taxon.

To make these definitions general, changes from the molecular level to that of adult physiology, behavior, biochemistry and morphology could be included within the context of "phenotypic." The important distinction to be made, however, is between the hierarchical level at which causal processes operate to establish innovations and those at which various processes might determine their historical fate. Given the conception of innovation implied by the preceding definitions, a mechanistic description of the origin of prospective innovations will lie within the domain of molecular biology and developmental genetics (e.g., Bonner 1982; Raff and Kaufman 1983; John and Miklos 1988), whereas a mechanistic description of the origin of evolutionary innovations will be found within the domain of population biology.

Most importantly, perhaps, once evolutionary innovations become established in a population so as to characterize it as a new evolutionary taxonomic unit, or species if you will, then, above that hierarchical level, patterns involving these innovations are interpretable as epiphenomena or effects. Arguably, there are no processes above the level of differentiating populations that determine the origin or establishment of evolutionary innovations, but there may be higher-level processes that determine the relative frequency of those innovations as expressed as patterns across space, time, and among taxonomic groups. To speak of macroevolutionary processes as those that account for large-scale innovations defining higher taxa (e.g., Alberch 1982:313; Gould 1982:340; Levinton 1988:2) misrepresents the hierarchical level at which processes producing innovation are located. Irrespective of whether innovations are perceived as "large" or "small," they all must arise and become established at the levels of individuals and populations, not higher taxa.

This discussion implies that the correct domain for studying innovation is between the molecular (processes of mutation and development) and populational (processes of differentiation) levels and that the approaches brought to bear on the problem of innovation will depend upon the questions asked and the focal level to which they are directed. Partitioning patterns and processes into their appropriate hierarchical levels also implies that the analysis of speciation will become increasingly important for the study of evolutionary innovations. It also suggests that a methodological protocol for the study of evolutionary innovation needs to be developed (see also Lauder and

Liem, in press).


## Speciation and Innovation: A Suggested Protocol

Evolutionary innovations are those fixed in basal evolutionary taxonomic units, and ideally it would be desirable to have a historical record of all the novelties – whether molecular, physiological, or morphological – manifested as differences among these taxa. Accordingly, a protocol for the historical analysis of evolutionary innovations will have the following four elements:

1. *Delineation of basal differentiated evolutionary units.* Processes of taxonomic differentiation result in populations that are diagnosably distinct from other such populations. Hypotheses about the boundaries of these basal taxa can therefore be formulated on the basis of a comparative analysis of discrete character variation, that is, of observable evolutionary innovations (Nelson and Platnick 1981; Cracraft 1983, 1987).

2. *Exhaustive sampling of evolutionary innovations.* At the level of basal taxa, diagnostic characters are evolutionary innovations. In the ideal case, we would like to have an exhaustive list of these characters for all basal taxa. In general, however, the kinds of innovations sampled and the extensiveness of the sampling effort will depend upon the questions being asked. Two types of characters must be sampled: those shared by basal taxa and those diagnostic of each basal taxon in the analysis.

3. *Erection of a phylogenetic hypothesis for the basal taxa.* A cladistic hypothesis will apportion postulated evolutionary innovations most parsimoniously across the tree. Indeed, without a phylogenetic hypothesis one cannot distinguish between innovations that are shared by two or more taxa as a result of common ancestry and those that apparently arose independently. If the history of innovations is to be interpreted correctly, these two types of pattern must be distinguished.

4. *Selection of clades with minimal extinction.* A complete record of

evolutionary innovations can be obtained only if the clade being studied has
not lost evolutionary taxa through extinction and if the investigator has sampled all the taxa. Extinction of one or more evolutionary taxa can lead to incorrect interpretations of character change, such as interpreting some characters as autapomorphies when in fact they are synapomorphies. This makes
other species appear more apomorphic than they really are.

Judging the relative amount of extinction within a group is a problematic exercise at best. In some cases, a good fossil record will inform whether
some taxa are extinct, but not all extinct taxa enter into the sedimentary record. And it is unlikely, particularly in terrestrial sedimentary regimes, for
the record to be sufficiently fine-grained across both space and time such
that all relevant evolutionary taxa will be preserved. The only other source
of data potentially capable of permitting inferences about extinction is that
of historical biogeography. If the clade under study has evolutionary taxa endemic in the same areas as taxa in other clades, and if these clades exhibit
biogeographic congruence with respect to the phylogenetic relationships of
their component taxa, then it may be possible to identify potential extinction
events when taxa in some clades are missing from certain of those areas.
Only two alternatives seem possible: either the taxon is present in the area
but unsampled or the taxon was never present. If one could make an ecological or biogeographic argument that a species should be expected in the area
in question but is lacking, then perhaps the case for extinction is strengthened. Realistically, however, it should be noted that conjectures about specific extinction events in the absence of direct evidence from the fossil record
border on the ad hoc and effectively mean one's interpretations must assume
some data to be missing. The use of biogeographic evidence may be appropriate when arguing that there has been no (or minimal) extinction as, for
example, when all areas of endemism are found to have an endemic species,
in which case all interpretations of character evolution are derived only from
known taxa. Despite our desire to know whether patterns of character distribution have been structured by extinction, precise knowledge of those influ-

ences will undoubtedly remain intractable for most groups of organisms.

This four-step protocol is a powerful tool with which to investigate evolutionary innovations. First, it helps identify innovations and establishes the hierarchical level at which they characterize one or more evolutionary taxa. This means that molecular and developmental biologists can use this historical hypothesis to discriminate between innovations and primitive retentions, and it sets the stage for interpreting developmental data historically.

Second, it permits the quantification of rates of change along different lineages. Thus, the magnitude of innovation in different evolutionary taxa can be estimated. Such information is essential if hypotheses of character evolution are to be tested. In particular, it allows us to ask whether large-scale changes in phenotype arise as evolutionary innovations, or whether most morphogenetic changes are of small phenotypic effect. More important, perhaps, it provides a method whereby the relative frequencies of these types of changes can be estimated.

Third, the method also permits the identification of historical patterns of covariance in characters. Given these patterns, many questions could be asked within the disciplines of developmental biology, functional morphology, and population biology.

**Speciation and Innovation: An Example.** In order to illustrate aspects of the above protocol, the pattern of speciation within a group of South American parrots will be examined. The genus *Pionopsitta* consists of at least eight basal evolutionary taxa, or phylogenetic species (Cracraft 1983; Cracraft and Prum 1988). They are distributed in nearly all of the major areas of endemism in tropical South America, and a hypothesis of their cladistic relationships is moderately corroborated using external characters (Cracraft and Prum 1988). Because there do not appear to be major distributional gaps, and because the area-relationships of the areas of endemism as specified by the cladistic relationships of these species are congruent with those of other clades of birds, a reasonable working hypothesis is that extinction in this group has been minimal.

The innovations (characters) to be discussed include external morpho-logical characters of coloration and color pattern (table 1), and an effort has been made to include all shared and unique characters. A phylogenetic hypothesis based on a cladistic analysis of these data (using the algorithm PAUP, Swofford 1985) is presented in figure 2. A single most parsimonious tree was found to have a length of 33 steps and a consistency index of 0.909 (there being one reversal and two cases of parallelism; see fig. 2).

This historical hypothesis forms a basis for a deeper understanding of evolutionary innovation within this clade. Some lineages and species are characterized by much more change than others. In particular, the sister

**PIONOPSITTA**

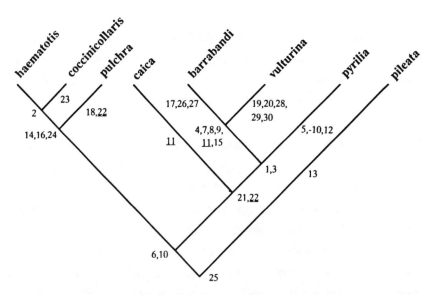

Figure 2. A phylogenetic hypothesis for the species in the South American parrot genus, *Pionopsitta* (after Cracraft and Prum 1988). Numbers refer to postulated derived characters ("innovations") listed in table 1; parallelisms are underlined and a negative sign indicates a reversal. Some lineages (e.g., the *barrabandi/vulturina* clade) exhibit higher rates of change, or changes with greater phenotypic effect, than their sister lineage. See text for discussion.

Table 1

Character-state data for the parrot genus *Pionopsitta* and its outgroup, *Hapalopsittaca*.

| | Characters and character-states* | | | | | | | | | | | | | | |
|---|---|---|---|---|---|---|---|---|---|---|---|---|---|---|---|
| Taxa | 1 | 2 | 3 | 4 | 5 | 6 | 7 | 8 | 9 | 10 | 11 | 12 | 13 | 14 | 15 |
| *H. melanotis* | 0 | 0 | 0 | 0 | 0 | 0 | 0 | 0 | 0 | 0 | 0 | 0 | 0 | 0 | 0 |
| *P. pileata* | 0 | 0 | 0 | 0 | 0 | 0 | 0 | 0 | 0 | 0 | 0 | 0 | 1 | 0 | 0 |
| *P. haematotis* | 0 | 1 | 0 | 0 | 0 | 1 | 0 | 0 | 0 | 1 | 0 | 0 | 0 | 1 | 0 |
| *P. coccinicollaris* | 0 | 1 | 0 | 0 | 0 | 1 | 0 | 0 | 0 | 1 | 0 | 0 | 0 | 1 | 0 |
| *P. pulchra* | 0 | 0 | 0 | 0 | 0 | 1 | 0 | 0 | 0 | 1 | 0 | 0 | 0 | 1 | 0 |
| *P. barrabandi* | 1 | 0 | 1 | 1 | 0 | 1 | 1 | 1 | 1 | 1 | 1 | 0 | 0 | 0 | 1 |
| *P. vulturina* | 1 | 0 | 1 | 1 | 0 | 1 | 1 | 1 | 1 | 1 | 1 | 0 | 0 | 0 | 1 |
| *P. pyrilia* | 1 | 0 | 1 | 0 | 1 | 1 | 0 | 0 | 0 | 0 | 0 | 1 | 0 | 0 | 0 |
| *P. caica* | 0 | 0 | 0 | 0 | 0 | 1 | 0 | 0 | 0 | 1 | 1 | 0 | 0 | 0 | 0 |

| | 16 | 17 | 18 | 19 | 20 | 21 | 22 | 23 | 24 | 25 | 26 | 27 | 28 | 29 | 30 |
|---|---|---|---|---|---|---|---|---|---|---|---|---|---|---|---|
| *H. melanotis* | 0 | 0 | 0 | 0 | 0 | 0 | 0 | 0 | 0 | 0 | 0 | 0 | 0 | 0 | 0 |
| *P. pileata* | 0 | 0 | 0 | 0 | 0 | 0 | 0 | 0 | 0 | 1 | 0 | 0 | 0 | 0 | 0 |
| *P. haematotis* | 1 | 0 | 0 | 0 | 0 | 0 | 0 | 0 | 1 | 1 | 0 | 0 | 0 | 0 | 0 |
| *P. coccinicollaris* | 1 | 0 | 0 | 0 | 0 | 0 | 0 | 1 | 1 | 1 | 0 | 0 | 0 | 0 | 0 |
| *P. pulchra* | 1 | 0 | 1 | 0 | 0 | 0 | 1 | 0 | 1 | 1 | 0 | 0 | 0 | 0 | 0 |
| *P. barrabandi* | 0 | 1 | 0 | 0 | 0 | 1 | 1 | 0 | 0 | 1 | 1 | 1 | 0 | 0 | 0 |
| *P. vulturina* | 0 | 0 | 0 | 1 | 1 | 1 | 1 | 0 | 0 | 1 | 0 | 0 | 1 | 1 | 1 |
| *P. pyrilia* | 0 | 0 | 0 | 0 | 0 | 1 | 1 | 0 | 0 | 1 | 0 | 0 | 0 | 0 | 0 |
| *P. caica* | 0 | 0 | 0 | 0 | 0 | 1 | 1 | 0 | 0 | 1 | 0 | 0 | 0 | 0 | 0 |

*Character (character-state) key (0=absent, 1=present): 1, extensive bright red wing lining; 2, extensive red on upper flank; 3, shoulder patch bright orange-red or yellow; 4, bright orange-red shoulder patch; 5, bright yellow shoulder patch; 6, upper breast yellow-green; 7, upper breast shining yellow-green; 8, belly deep blue-green; 9, tibial feathering with extensive yellow, orange, or red; 10, crown and hindneck darkish, with moderate to heavy melanin deposition; 11, crown and back of neck black; 12, crown, throat, and back of neck yellow; 13, crown bright red in male, not female (sexual dimorphism); 14, crown and upper back dirty yellow-green; 15, throat with extensive deep black; 16, red on malar and auricular; 17, yellow malar patches; 18, face extensively pinkish red; 19, head feathers reduced; 20, yellow collar around head; 21, basal portion of inner web of tail yellow (ventral view); 22, trailing edge of primaries green (ventral view); 23, breast with red; 24, red on tail (ventral view); 25, lores only partly feathered; 26, yellow-orange shoulder patch; 27, yellow tibial feathering; 28, red shoulder patch; 29, wing lining bright scarlet; 30, red tibial feathering.

species *barrabandi* and *vulturina* are markedly distinct morphologically relative to the other species, and each has diverged significantly from their common ancestry. Indeed, *vulturina* has sometimes been placed in a monotypic genus.

These findings raise interesting questions for workers focusing at lower hierarchical levels. If it is assumed that all taxa of this clade have been sampled and there has been no extinction, then the relatively higher frequency of innovations in the *barrabandi-vulturina* lineage and in *vulturina* itself suggests that either the frequency of mutational events of small effect has increased or some mutational events have had greater phenotypic effect than in other parts of this clade. In the same vein, if it is assumed that rates of mutational events were approximately equal across all lineages, then events having a marked influence on morphogenetic pathways affecting plumage pattern and coloration were about as frequent as those having small effects (compare, for example, the numbers of innovations along each lineage).

If more were known about the developmental genetics of these characters, it might be possible to substitute strong inference for speculation. Still, examination of the patterns of covariation seems to imply that there might have been single mutational events that caused moderately extensive modification of the phenotype by perturbating morphogenetic pathways affecting the deposition of various pigments at different locations on the body. For example, two of the characters (11 and 15 of table 1) shared by *barrabandi* and *vulturina* involve deposition of melanin in feathers of the head; in *barrabandi* all three characters (17, 26, 27) include a change in carotenoid pigmentation to express yellow coloration at three different locations; and in *vulturina* a modification in carotenoid pigmentation is expressed as red coloration at several of the same locations as the changes in *barrabandi*.

The message of this example is that a protocol involving the cladistic analysis of differentiated populations, or basal taxa, is required in order to recognize innovations and define their historical pattern (see also Larson et al. 1981). Moreover, the protocol permits us to identify sister taxa and to investigate historically how alterations in morphogenetic pathways might relate causally to adult differences. Only one set of characters has been sampled in this example, and others might yield different patterns. Nevertheless, the example itself is realistic in the sense that an exhaustive enumeration of all evolutionary novelties is beyond the scope of any analysis.

### Key Innovations and the Causal Analysis of Diversification

Patterns of taxonomic diversification have generally been explained within an adaptationist context (e.g., Simpson 1953; Mayr 1963), and a prominent component of such explanations is the key innovation: "In general, adaptive radiations will not occur until after an evolutionary novelty has reached a certain degree of development . . ." (Liem 1973:426). This claim is first one of correlation and then of causality. A group of high diversity is identified and some diagnostic character or character-complex is noted and proposed to be a key innovation. By baptizing this character a *key* innovation, the biologist conjectures a causal relationship between the appearance of that innovation and the subsequent diversification of the group. It is worthwhile to scrutinize this claim in more detail.

Changes in diversity within a monophyletic group are the result of changes in rates of speciation and/or extinction. With respect to increases in diversity, the rate of speciation must increase relative to the rate of extinction, which can come about as the result of an increase in the rate of speciation or a decrease in the rate of extinction, or both.

Three broad categories of factors are potentially responsible for increasing the rate of speciation. The first includes those factors that increase the rate at which populations become isolated. The higher this rate, all things being equal, the higher should be the rate of differentiation. For example, an increase in geomorphological complexity would tend to increase the rate of vicariance and thus allopatry (Ross 1972; Cracraft 1985).

A second set of factors includes those that increase the rate at which novelties enter populations. These factors reside at the level of the genome and are mutational events that alter developmental pathways. Again, all things being equal, the higher the frequency of mutational events, the more variability within populations and the more likely some of that variation will become fixed.

The final set of factors includes those increasing the rate at which novelties become fixed in populations. This might entail intensification of directional natural selection or sexual selection, or it could include increases in fixation because of the influences of stochastic events (relative differences in population sizes, for instance).

How might the key innovations listed above have an influence on speciation rate? It is clear that the majority cannot. First, most lack ontological status as true evolutionary novelties. Most are composites of numerous true evolutionary novelties and are often artifacts of extinction and sampling, and artifacts cannot be causal of anything. Secondly, even if one of these innovations were a singular evolutionary novelty, the real problem would be to discover how that innovation could have increased the speciation rate of the clade possessing it as a synapomorphy. Such innovations must influence speciation through one of the three factors just mentioned. Rigorous causal hypotheses may be exceedingly difficult to formulate because they will have to be based on something other than the correlation between cladal diversity and cladal synapomorphy. Unique events, moreover, do not lend themselves to rigorous causal analysis (Lauder 1981; Stiassny and Jensen 1987; Raikow 1986, 1988): "Suppose we agree that wings are a key adaptation of bats. How can we show that they are responsible for there being ca. 870 species as against, say, 87 or 8,700?" (Raikow 1988:77). Indeed, how might we demonstrate that a "key innovation," which is present in a speciose group but not in its much less diverse sister taxon, is responsible for elevating speciation rate in the former? Three answers to this question have been suggested:

1. *The comparative argument.* It is claimed that causation can be inferred if we find the same novelty in unrelated taxa, all of which exhibit high diversity. Kochmer and Wagner (1988) have suggested, for example, that small body size is in some way causative of high diversity because the species of many speciose groups are comprised of small individuals. Ignoring some obvious difficulties of "small size" as a key innovation, arguments of this type identify not causation but multiple correlation. To confound matters, moreover, such "correlations" are almost always spurious because diverse "groups" are chosen a posteriori (the taxonomic artifact noted by Raikow 1988) and numerous counterexamples are inevitably ignored (for example, there are countless clades of low diversity in which individuals are of small body size: were these also used to document the "correlation"?). Even if one restricted comparisons to sister clades, it is not immediately obvious how the sample universe could be chosen without bias, short of including all sister clades, an impossible task since phylogenetic relationships for virtually all organisms are so poorly resolved. Choosing taxa of a particular rank, such as families or orders, as Raikow (1988:78) has noted, is arbitrary and thus imposes an artificiality on the data in the sense that such a procedure excludes all other clades from the sample: why choose a clade for comparison simply because it has historically been called a family or order? Why, moreover, should taxa of the same rank be considered equivalent, or comparable, across groups, or even within a relatively closely related group such as birds or mammals (see Cracraft 1981)?

2. *The functional argument.* Causation is inferred, first, from a correlation between some novelty and high diversity, and, second, from a conjecture that the novelty is functionally, or adaptively, important. Liem (1973) proposed, for example, that a few simple modifications of the jaw apparatus of cichlid fishes comprised a key innovation that enabled cichlids to specialize on a variety of food types in a variety of habitats and thus attain high diversity. Arguments formulated in these terms, however, are little more than restatements of the original correlation: the family is diverse and we

observe that its species are also structurally and ecologically diverse.  A causal connection to speciation and extinction rates still has not been demonstrated.

The functional argument takes on the appearance of demonstrating the causal importance of key innovations when the novelty is said to be directly related to speciation or extinction rate control.  One such study said to "test" (Fitzpatrick 1988) the hypothesis of a causal relationship between an innovation and speciation rate is Ryan's analysis (1986) of anuran diversity. Ryan found a correlation between complexity of an inner ear organ (the amphibian papilla) and cladal diversity.  Physiological data indicate that the more complex the amphibian papilla, the greater is its sensitivity to a range of frequencies, and Ryan reasoned that "since mating call divergence is an important component of the speciation process, differences in the number of species in each lineage should be influenced by structural variation of the inner ear" (Ryan 1986:1379).  The most diverse families, he observed, possess the most structually complex form of the amphibian papilla.

It is questionable whether these kinds of data constitute a test of the hypothesis of causation.  First, the link to causation basically restates the original observation:  the most diverse taxa exhibit the most variation in vocal frequencies and have the most complex amphibian papilla.  Because all diverse clades ("families" in this case) have the same form of the amphibian papilla, its mere presence does not directly tell us how the taxonomic diversity and its broad range of vocal behavior was generated.  Second, the causal correlation may well be spurious because it is merely a synapomorphy of a diverse clade.  If this character is truly causal of high diversity, one might predict that sister taxa within this large clade would have nearly equal diversity.  In fact they do not.  Some sister taxa having a complex amphibian papilla differ in diversity by as much as an order of magnitude (e.g., Leptodactylidae versus Dendrobatidae).  If variation in the structure of the amphibian papilla is causally related to an increased speciation rate, such an inference cannot emerge in a simple fashion from observed correlations to cladal diversity. The fact that these sister taxa differ in diversity demonstrates that other factors underlie variation in diversity, and if other factors are operable, how do we specify – and we must if the hypothesis of causation is truly "testable" – the contribution made by the amphibian papilla? In this case at least, it seems that the hypothesis of causation is not "testable" beyond the original correlation.

3. *The "necessary but not sufficient" argument*. Ryan's example of the amphibian ear is but one of a class of "key innovations" which might be termed instances of the "necessary but not sufficient" argument for the role of innovations in diversification. Other examples include the cichlid jaw apparatus (Liem 1973), gene duplications (Lauder and Liem, in press), and an increase in biomechanical complexity between the ray-finned and halecostome fishes (Lauder and Liem, in press).

The argument is usually structured as follows: evolutionary innovations are necessary but by themselves are not sufficient to account for differences in speciation or extinction rate. A new key innovation is said to create the potentiality for evolutionary change to occur along multiple morphogenetic pathways whereas others preclude such modifications. Complex structural systems are more susceptible to evolutionary change than those under strong functional/developmental constraint. If a change can increase structural complexity, it increases the chances of, and even facilitates, evolutionary differentiation. Thus, the high diversity of cichlid fishes and more advanced anurans might not have been possible without their respective innovations. As Liem (1973) and Ryan (1986) suggest, the cichlid jaw apparatus and the advanced amphibian papilla might have increased the potential for variability in feeding and vocal behavior, respectively. Schaefer and Lauder (1986:504) have expressed this argument as their "decoupling hypothesis": "a phylogenetic increase in the number of biomechanical components (as by the decoupling of primitively constrained elements) in a morphological system is related to morphological and functional diversity because of the increased possibility for change and novel connections between independent components in a complex system. An increase in constructional flexibility is also expected to correlate with the acquisition of new functions."

How might the "necessary but not sufficient" argument be evaluated? First, the examples themselves are poorly defined as evolutionary innovations (see earlier discussion). The "amphibian papilla" and "cichlid jaw apparatus" are structural patterns seen between higher taxa, not discrete innovations postulated to have arisen at the level of differentiating populations. If, in fact, these "key innovations" arose as numerous sequential modifications ("innovations") associated with many speciation events, how might this influence our perception that these "key innovations" were necessary for subsequent diversification? Here the ontological problem of individuating a "key innovation" is critical for dissecting causation. If these features are not true evolutionary innovations arising at the population level, but are instead composites of numerous innovations, what justification do we have for ascribing a causal role to something that is not a true innovation?

Even admitting the possibility that these are true evolutionary innovations does not explain why and how the resultant variation emerged over evolutionary time in the guise of taxonomic diversity. Variation in diversity among clades of cichlids or anurans cannot be attributed to the presence of a particular jaw mechanism or inner ear structure since each is a shared synapomorphy. A single innovation does not, in general, make a species "more complex" relative to its sister species. Such complexity arises as a result of sustained diversification: characters are modifications of other characters. Without diversification, there would be no complexity to explain: complexity is an effect of diversification, not its cause. Therefore, a deterministic explanation of patterns of diversity must lie elsewhere. If a character said to be a "key innovation" cannot explain the patterns of diversity among the clades possessing it, then it must be concluded that this feature cannot also be singled out as the reason for high diversity relative to a less diverse sister group.

What about extinction rates? Can innovations influence those? Here, too, the conceptual and empirical waters are murky. Species, *as cohesive en-*

*tities or units*, do not become extinct. Species extinction is a cumulative effect of the deaths of individual organisms. Consequently, any argument that an evolutionary innovation is instrumental in influencing extinction rate must necessarily focus at the level of individual organisms and populations. How might "key innovations" such as those mentioned earlier affect the fecundity or longevity of individual organisms and thus have an indirect effect at the level of the species? Probably not at all. Another question is whether innovations present at the level of the species can be passed on to descendant species (Cracraft, in press). It is not at all clear, for example, that species have emergent properties for extinction resistance; these can generally be interpreted as an effect of the properties of individual organisms that may be manifested at higher levels. Once again, it probably will prove very difficult to erect a testable hypothesis claiming a relationship between a particular innovation and an influence on extinction rate within a clade.

In conclusion, the concept of key innovation seems to have originated from the desire to extend an adaptationist form of interpretation to hierarchical levels higher than the one at which Darwinian natural selection and adaptation operate. Thus, one can be an ardent adaptationist and still reject the utility of the concept of key innovation. The ontological, methodological, and empirical themes discussed in this chapter suggest that the concept of key innovation has little to tell us about the causal dynamics of patterns of biological diversification such as those expressed among clades.

### Acknowledgments

I wish to thank Drs. Allan Larson, George V. Lauder, Robert J. Raikow, and David B. Wake for their extremely helpful and critical comments on the manuscript. This work was supported by a grant (BSR-8520005) from the National Science Foundation.

## References

Alberch, P. 1982. Developmental constraints in evolutionary processes, pp. 313-32. In *Evolution and Development*, ed. J. T. Bonner. Berlin: Springer-Verlag.

Bonner, J. T., ed. 1982. *Evolution and Development*. Berlin: Springer-Verlag.

Cracraft, J. 1981. Pattern and process in paleobiology: The role of cladistic analysis in systematic paleontology. *Paleobiology* 7: 456-68.

Cracraft, J. 1983. Species concepts and speciation analysis. *Current Ornithology* 1: 159-87.

Cracraft, J. 1985. Biological diversification and its causes. *Annals of the Missouri Botanical Garden* 72: 794-822.

Cracraft, J. 1986. The origin and early diversification of birds. *Paleobiology* 12: 383-99.

Cracraft, J. 1987. Species concepts and the ontology of evolution. *Biology and Philosophy* 2: 63-80.

Cracraft, J. 1988. The major clades of birds, pp. 339-61. In *The Phylogeny and Classification of the Tetrapods*, Vol. 1, ed. M. J. Benton. Oxford: Clarendon Press.

Cracraft, J. In press. Species as entities of biological theory. In *What the Philosophy of Biology Is*, ed. M. Ruse. Dordrecht: Kluwer Academic Publishers.

Cracraft, J., and R. O. Prum. 1988. Patterns and processes of diversification: Speciation and historical congruence in some Neotropical birds. *Evolution* 42: 603-20.

Fitzpatrick, J. W. 1988. Why so many passerine birds? A response to Raikow. *Systematic Zoology* 37: 71-76.

Futuyma, D. J. 1986. *Evolutionary Biology*. Sunderland, MA: Sinauer Associates.

Gauthier, J. 1986. Saurischian monophyly and the origin of birds. *California Academy of Sciences Memoirs* 8: 1-55.

Gauthier, J., and K. Padian. 1985. Phylogenetic, functional, and aerodynamic analyses of the origin of birds and their flight, pp. 185-97. In *The Beginnings of Birds*, ed. M. K. Hecht, J. H. Ostrom, G. Viohl, and P. Wellnhofer. Eichstatt, W. Germany: Freunde des Jura-Museums.

Gould, S. J. 1982. Change in developmental timing as a mechanism of macroevolution, pp. 333-46. In *Evolution and Development*, ed. J. T. Bonner. Berlin: Springer-Verlag.

John, B., and G. Miklos. 1988. *The Eukaryote Genome in Development and Evolution*. London: Allen & Unwin.

Kochmer, J. P., and R. H. Wagner. 1988. Why are there so many kinds of passerine birds? Because they are small. A reply to Raikow. *Systematic Zoology* 37: 68-69.

Larson, A., D. B. Wake, L. R. Maxson, and R. Highton. 1981. A molecular phylogenetic perspective on the origins of morphological novelties in the salamanders of the tribe Plethodontini (Amphibia, Plethodontidae). *Evolution* 35: 405-22.

Lauder, G. V. 1981. Form and function: Structural analysis in evolutionary morphology. *Paleobiology* 7: 430-42.

Lauder, G. V., and K. F. Liem. In press. The role of historical factors in the evolution of complex organismal functions. Dahlem Conference, 1988.

Levinton, J. 1988. *Genetics, Paleontology, and Macroevolution.* New York: Cambridge University Press.

Liem, K. F. 1973. Evolutionary strategies and morphological innovations: Cichlid pharyngeal jaws. *Systematic Zoology* 22: 425-41.

Martin, L. D. 1985. The relationships of *Archaeopteryx* to other birds, pp. 177-83. In *The Beginnings of Birds*, ed. M. K. Hecht, J. H. Ostrom, G. Viohl, and P. Wellnhofer. Eichstatt, W. Germany: Freunde des Jura-Museums.

Mayr, E. 1963. *Animal Species and Evolution.* Cambridge: Harvard University Press.

Nelson, G. J., and N. I. Platnick. 1981. *Systematics and Biogeography: Cladistics and Vicariance.* New York: Columbia University Press.

Ostrom, J. H. 1976. *Archaeopteryx* and the origin of birds. *Biological Journal of the Linnean Society* 8: 91-182.

Padian, K. 1982. Macroevolution and the origin of major adaptations: Vertebrate flight as a paradigm for the analysis of patterns. *Third North American Paleontogical Convention Proceedings* 2: 387-92.

Raff, R. A., and T. C. Kaufman. 1983. *Embryos, Genes, and Evolution.* New York: Macmillan.

Raikow, R. J. 1986. Why are there so many kinds of passerine birds? *Systematic Zoology* 35: 255-59.

Raikow, R. J. 1988. The analysis of evolutionary success. *Systematic Zoology* 37: 76-79.

Ross, H. H. 1972. The origin of species diversity in ecological communities. *Taxon* 21: 253-59.

Ryan, M. J. 1986. Neuroanatomy influences speciation rates among anurans. *Proceedings of the National Academy of Sciences* 83: 1379-82.

Sanz, J. L., J. F. Bonaparte, and A. Lacasa. 1988. Unusual Early Cretaceous birds from Spain. *Nature* 331: 433-35.

Schaefer, S. A., and G. V. Lauder. 1986. Historical transformation of functional design: Evolutionary morphology of feeding mechanisms in loricarioid catfishes. *Systematic Zoology* 35: 489-508.

Simpson, G. G. 1953. *The Major Features of Evolution.* New York: Columbia University Press.

Stiassny, M. L. J., and J. S. Jensen. 1987. Labroid intrarelationships revisited: Morphological complexity, key innovations, and the study of comparative diversity. *Bulletin of the Museum of Comparative Zoology* 151: 269-319.

Swofford, D. L. 1985. *PAUP, Phylogenetic Analysis Using Parsimony*, version 2.4. Champaign: Illinois Natural History Survey.

# GENETICS AND DEVELOPMENT

# The Evolutionary Genetics
# of Adaptation

*Brian Charlesworth*

As Charles Darwin clearly recognized, the evolution of a complex structure or behavior pattern requires the coordination of numerous different component parts, which must be mutually adjusted if the whole is to function efficiently. This is true even of the simplest examples of adaptation, other than the classic cases of single-character change such as industrial melanism or resistance to poisons (Bishop and Cook 1981), which are under simple genetic control. Mimicry in butterflies or heterostyly in plants provide examples of variations within single species where several, relatively simple, characters are coordinated to produce functional structures of adaptive importance (see fig. 1). More complicated adaptations, such as the vertebrate eye, simply represent an elaboration of this principle.

Darwin also recognized that the evolution of such character complexes posed a considerable intellectual challenge to his theory of evolution by natural selection. His answer to this challenge constitutes what I shall call the *class 1* explanation of the evolution of adaptation: a complex adaptation arises from the successive incorporation of changes to the component parts, each of which is advantageous at the current state of the system. In the twentieth century, this view is perhaps most strongly associated with R. A. Fisher and his followers (Fisher 1930).

An alternative theory was vigorously advocated by Richard Goldschmidt (1940) but was originally proposed by early Mendelians such as de Vries (Provine 1971). On this view, a single mutation is capable of produc-

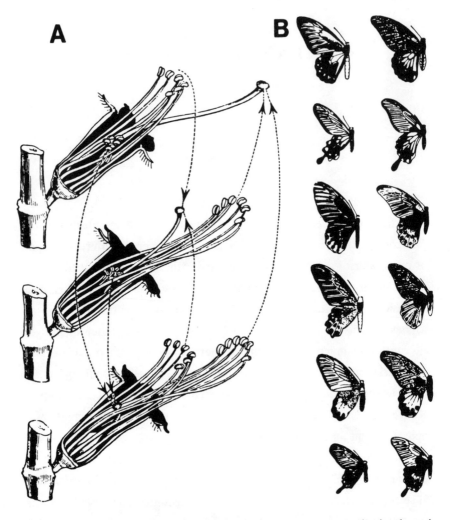

Figure 1. A. The three morphs of the tristylous species *Lythrum salicaria,* showing the positions of the stamens and anthers. The dashed lines indicate the pollinations which are permitted by the incompatibility mechanism: pollen from a given anther position is compatible only with a stigma at the same level as itself. Modified from Darwin 1865, fig. 1. B. Six forms of the Batesian mimic *Papilio memnon* (right) and the corresponding models (left). Modified from Turner 1984, fig. 15.8.

ing a coordinated set of changes in a character complex. Fixation of such a mutation was supposed to cause a saltatory evolutionary step, creating a new structure in one blow. According to Goldschmidt, this mode of evolution is responsible for the origination of species and higher taxa, and is to be distinguished from the essentially Darwinian processes of evolution at lower taxonomic levels. I shall refer to such a process of saltational evolution as the *class 2* explanation. Something close to this interpretation has been revived recently in the writings of Løvtrup (1974), Gould (1980, 1982), Stanley (1979) and Alberch (1980). Gould (1982), for example, writes:

> If new Bauplane often arise in an adaptive cascade following the saltational origin of a key feature, then part of the process is sequential and adaptive, and therefore Darwinian but the initial step is not, since selection does not play a creative role in building the key feature.

The *class 3* explanation appeals to the effects of random genetic drift in causing a chance shift from one equilibrium under natural selection to another equilibrium (Haldane 1932; Wright 1932, 1980). Using Wright's metaphor of selective peaks (fig. 2), drift can cause a stochastic transition across an adaptive valley separating two peaks, enabling the population to reach a state which could never be attained under the influence of selection alone. As discussed by Haldane (1932:102), this mechanism provides a possible solution to the problem of evolving a new adaptation when the initial steps are selectively disadvantageous.

In this chapter, I shall first consider the theoretical arguments for and against each of these classes of explanation of adaptations. I will then discuss evidence from genetic studies that help to discriminate among them. My conclusion is that there is strong theoretical and empirical evidence against the class 2 explanation. While there is no direct empirical evidence that excludes class 3, it is far more sensitive to population structure than class 1. Furthermore, class 1 makes several detailed predictions concerning a number of cases of adaptation that have been well studied genetically, and which are verified empirically. No evidence that compels us to postulate the operation of a class 3 process for the evolution of an adaptation is currently available. I therefore conclude that the Darwinian explanation of adaptation is best supported by the evidence.

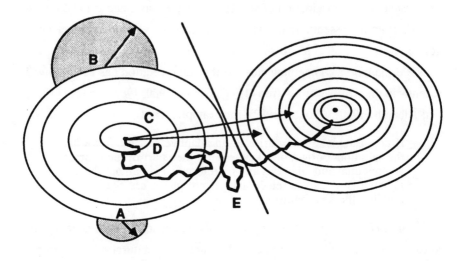

Figure 2. Possible modes of evolution in a two-dimensional adaptive landscape with two selective peaks. The elliptical curves indicate contours of fitness as functions of the two variables under selection, with fitness increasing toward the center of each set. The peak on the right has a higher fitness value, as indicated by the greater density of contour lines. The dotted areas enclosed in the semicircles around the arrows connected to points A and B show the probability of a random change in the character causing a *decrease* in fitness, for a small (A) and a large (B) change. The arrows connected to points C and D show that a large change (C) can cause a shift toward higher fitness, as a result of crossing the adaptive valley between the two peaks (the oblique line), whereas a small change (D) does not. The heavy curve (E) shows a possible path toward the higher peak that may be traversed by a population as a result of random genetic drift.

## Theoretical Considerations

**Class 1 Explanations.** The theoretical basis of this mechanism is straightforward. The classic results of theoretical population genetics show that natural selection operating on Mendelian variation is capable of overcoming the effects of mutation or random genetic drift, except in cases where the pressure of selection is very weak and the effective population size is very small (Fisher 1930; Wright 1931; Haldane 1932). At this time, it is difficult for us to appreciate the importance of this conclusion for evolutionary biology. It removes all of the relatively well-founded objections to natural selection, such as the decay of variability under blending inheritance, and hence eliminates the need to appeal to the inheritance of acquired characteristics as a supplementary force in evolution that was so strongly felt by Darwin (Fisher 1930, chap. 1).

Given that natural selection acting on Mendelian variation is a mechanistically credible evolutionary force, can it produce the evolution of a complex adaptation in the sense defined above? Fisher (1930, chap. 2) argued that such evolution must proceed by a series of small steps. He constructed a model of a multidimensional character set, with a single point corresponding to maximum fitness as a function of the vector of character values (fig. 2). An infinitesimally small change to a point located some way from the optimum has approximately a 50% chance of carrying the system closer to the optimum, whereas a large change will have a high chance of carrying it *away* from the optimum. Fisher argued that this means that the evolution of a complex character necessarily requires the accumulation of minor changes, and is overwhelmingly unlikely to involve large changes.

This argument certainly carries a lot of weight. It is most appropriate when an environmental change creates a situation where the species finds itself located some distance from the new optimum state. This must be a very frequent mode of adaptive evolution, given the association between occupation of new ecological niches and the evolution of new adaptive modes (Wright 1949; Simpson 1953). It is possible, however, that the environment remains constant, but there are two (or more) selective peaks separated by a valley (Haldane 1932:102; Wright 1932). A selective transition from one peak to a point of higher fitness near the second peak is clearly impossible by small steps (fig. 2), but a change of sufficiently large size may have this effect. This requires, of course, that the second peak be associated with substantially higher fitness. Furthermore, the larger the number of dimensions, the smaller the chance of moving in the direction of the valley between two peaks by a random perturbation (it is easily seen to be approximately $1/(2n)$, where $n$ is the number of dimensions). Thus, this mode of evolution is most likely in situations where the adaptation involves a relatively small number of component characters. The initial major change will probably carry the system into the neighborhood of the second peak. Completion of the process will require a number of minor steps on the lines envisaged in Fisher's model. This "two-step" process was apparently first proposed by Nicholson (1927) and Fisher (1930, chap. 7) in relation to mimicry. It has subsequently been elaborated in this context by Sheppard (1962), Charlesworth and Charlesworth (1976a, 1976b), and Turner (1977, 1984).

**Class 2 Explanations.** Modern advocates of this mode of evolution emphasize the importance of developmental mechanisms in constraining the kinds of phenotype that can be achieved by members of a given species, and suggest that such constraints can be overcome by the effects of mutations with sufficiently profound effects (Løvtrup 1974; Stanley 1979; Gould 1980; Alberch 1980). The existence of variation between and within species in discrete characters, such as numbers of digit elements in salamanders (Alberch 1980), is often cited as evidence for the plausibility of this kind of process. However, this mechanism suffers from several difficulties. Its most notable problem is that it is hard to see how a functional complex of characters could be put together in one step. The considerations discussed above show that the chance of such an event is very small for a multidimensional character complex, assuming mutation in a random direction. Furthermore, as noted by Fisher (1930, chap. 1), major mutations almost always have pleiotropic consequences with adverse effects on fitness. In random-mating populations, autosomal mutations have a significant chance of fixation by selection only if they are expressed when heterozygous (Haldane 1924, 1927). Nearly all mutations in *Drosophila* and mammals that have major morphological effects when heterozygous are recessive lethal, as a result of pleiotropic side effects (Hadorn 1961). Hence, such major mutations are at best merely likely to become polymorphic, rather than fixed. Furthermore, naturally occurring variation in discrete characters is usually found to have an underlying basis of polygenic inheritance coupled with thresholds (Wright

1968). Hence, no inferences can be drawn from purely phenotypic evidence concerning the genetic mechanisms of evolutionary change in such characters (Charlesworth et al. 1982).

A possible answer to the objection that a well-integrated set of characters is unlikely to arise in one step is that developmental regulation may cause the automatic readjustment of other body parts in response to the perturbation of a subset of structures (Goldschmidt 1940; Thomson 1987). An example of such a situation is provided by the goat described by Slijper, in which the absence of the forelimbs was compensated for by the development of an unusual arrangement of muscles, ligaments, and bones, enabling the animal to hop like a kangaroo (Maynard Smith 1975:301). However, this creature did not produce offspring, and so its potential as a "hopeful monster" is purely speculative. The evidence from genetics is overwhelmingly unfavorable to the idea that single Mendelian genes with major morphological effects on a complex of characters can become established in evolution (see the discussion of the empirical evidence below).

Indeed, Goldschmidt postulated a special class of mutation, the systemic mutations, in which a radical repatterning of chromosome structure supposedly led to a repatterning of development affecting whole character complexes in one step. He viewed this as a means of leaping the "unbridgeable gap" that he perceived between intraspecific evolution, and differences between species and higher taxonomic levels. A more modest role for chromosomal rearrangements as agents of morphological evolution has recently been proposed by Allan Wilson and coworkers (Wilson et al. 1975; Bush et al. 1977), who suggest that such rearrangements may cause regulatory mutations with favorable effects on fitness when homozygous. But rearrangements generally reduce fertility when heterozygous, due to the production of aneuploid or duplication/deficient gametes with lethal effects on the zygotes that carry them. The fixation of a rearrangement thus requires genetic drift to cause an increase in its frequency past the equilibrium point generated by the heterozygote disadvantage with respect to net fitness produced in this way, and so it corresponds more properly to the class 3 model discussed below. Detailed theoretical analyses of this process show that rearrangements with sizeable effects on fitness are extremely unlikely to become fixed, except in species where local population size is small, migration is limited, and there is a good deal of turnover of populations by local extinction and recolonizations (Wright 1941; Lande 1979, 1985a; Walsh 1982).

**Class 3 Explanations.** The classic development of this model of adaptive evolution is Sewall Wright's shifting-balance theory (Wright 1932, 1980). This postulates a species divided into a large number of local populations of relatively small size, and between which migration is limited but not absent. Random genetic drift causes perturbations from the original equilibrium under selection (selective peak). A population which drifts across an adaptive valley to a peak associated with higher mean fitness (fig. 2) will increase somewhat in numbers, and hence send out more migrants to other populations than the average. In this way, a selectively superior gene or character combination produced by drift may spread through the species as a result of differential migration, causing the entire species to shift to a new, superior, selective peak in the absence of any environmental change.

Wright himself did not conduct any detailed examination of the rate of evolution that could be expected under the above scenario. The studies of the fixation of chromosome rearrangements referred to above represent the simplest realization of the model. More recently, a study of a nongenetic model of evolution involving stochastic peak shifts in a single population was performed by Newman et al. (1985). Lande (1985b, 1986) has analyzed a model of a stochastic peak shift involving a quantitative character in a single population (fig. 2). This study shows that there is an extremely long waiting time to a stochastic jump from one equilibrium to the other, with the intensity of selection that one might normally expect for a morphological character, unless the population size is very small.

The only study of selection on a quantitative character that deals directly with the problem of the probability of a stochastic peak shift arising in a single local population and ultimately spreading through an entire species is that of Rouhani and Barton (1987b). They conclude that such an event will be extremely infrequent, except in a species where local population size is very small, permitting a high chance of a stochastic shift in local populations. For two-dimensional arrays of populations, they reach the interesting conclusion that the rate of evolution is nearly independent of the strength of selection.

Clearly, all of these studies are limited to essentially one-dimensional characters. Intuitively, one feels that the chance of a transition will be lower, the larger the number of dimensions, if the characters concerned are relatively independent genetically. This is because there are more directions in which to move by chance, so that there is a smaller chance of movement in the direction of the adjacent peak (see discussion of mechanism 1 above), but this problem requires further theoretical study.

A somewhat different mode of evolution by stochastic shifts involves founder events, in which a small population becomes isolated from the parental population, and grows rapidly back to a large size. Several authors have suggested that random genetic drift during the generations when the population size is low may cause a stochastic transition between alternative selective equilibria (Mayr 1954, 1963; Carson 1975, 1982; Templeton 1980; Carson and Templeton 1984). In the present context, the most relevant type of event of this kind involves a quantitative character with two or more optima under selection, of the type discussed above (fig. 2). Models of this situation have been analyzed by Rouhani and Barton (1987a) and Charlesworth and Rouhani (1988), who used a univariate metrical character subject to a double-peaked selection function. They conclude that, while shifts in mean phenotype of a substantial magnitude can occur by this process, their probability of occurrence is extremely low except when recovery from the bottleneck is sufficiently slow that a very large reduction in heterozygosity at neutral loci would be expected. There is no evidence for such a reduction in heterozygosity at enzyme loci in groups such as the Hawaiian *Drosophila,* which are premier candidates for founder effect speciation (Barton and Charlesworth 1984; Carson and Templeton 1984).

### Empirical Evidence

**Intraspecific Variation.** Mimicry in butterflies has perhaps supplied the richest source of genetic data concerning the evolution of adaptations (Sheppard 1959, 1962; Turner 1977, 1984; Sheppard et al. 1985). Heterostyly in flowering plants is the closest rival system (Barrett 1989), but will not be discussed here. Two major classes of mimicry have been studied: Müllerian mimicry, where two distasteful species share similar warning patterns, and Batesian mimicry, where an edible species acquires a pattern similar to that of a warningly colored, distasteful species. In many cases the degree of resemblance is extraordinarily detailed, and involves several different aspects of pattern and coloration (fig. 1).

The theoretical analysis of the dynamics of selection on Müllerian mimicry have suggested that a two-step process is most likely under class 1 models. This involves selection for an initial major gene mutation conferring a substantial but imperfect degree of resemblance, followed by more minor adjustments leading to perfection of the resemblance (Turner 1984; Sheppard et al. 1985). This is because each distasteful species, initially very different in pattern, is subject to stabilizing selection on pattern, since minor deviants from the prevailing pattern for a species will not be recognized by the predator, and hence will lose the "safety-in-numbers" advantage characteristic of Müllerian mimicry (fig. 3). A sufficiently large change can carry the less protected species into the zone of protection of the better protected species, and thus can confer a selective advantage on its carriers.

The numerous geographical races of the South American butterflies *Heliconius erato* and *H. melpomene* exhibit very different patterns, but within

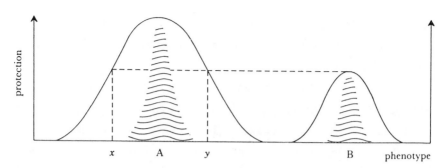

Figure 3. The curve of protection against phenotype for two distasteful model species, A being more abundant than B. The wavy lines indicate the distribution of phenotypes within each species. A mutation that causes a member of species B to acquire a phenotype in the range x to y will enjoy greater protection, but any other change will result in a loss in fitness. Once B has moved to the left of y, smaller changes that bring it closer to the peak for A will be selected for. Modified from Sheppard et al. 1985, fig. 11.

a given area the two species behave as almost perfect Müllerian mimics. This has provided a wealth of material for genetic analysis of the evolutionary origin of the different patterns. The results of these studies strongly agree with the predictions of the above model (Turner 1984; Sheppard et al. 1985). There are many different loci involved in control of the wing patterns of these species. Reconstruction of the phylogenies of the geographic races shows that each transition between successive states has involved a small number of genetic changes of a relatively major kind, each involving a single element of the character complex, together with modifier genes whose effects are too small to be individually analyzable. The complicated differences in pattern that can be seen between present-day races are the result of numerous changes, each involving individual elements of the patterns.

Similarly, genetic studies of Batesian mimicry in the genus *Papilio* provide strong support for class 1 models, and conclusively refute class 2 models (Sheppard 1959; Clarke and Sheppard 1960; Charlesworth and Charlesworth 1976b; Turner 1977, 1984). In contrast to Müllerian mimicry, selection on a mimetic variant of an edible species that resembles a distasteful model species is negatively frequency-dependent (Fisher 1930, chap. 7), since a high frequency of mimics relative to models leads to a failure of the predators to associate the model's pattern with distastefulness, and hence avoid insects that resemble the model. This results in extensive polymorphism in Batesian mimics, often involving several mimetic forms, each resembling a quite different species of model. It was early recognized that these different forms of various *Papilio* species segregate as alleles at a single genetic locus. This led Punnett (1915) and Goldschmidt (1945) to argue that the complex patterns distinguishing these forms had each arisen by a single mutational step.

Conversely, Nicholson (1927) and Fisher (1930, chap. 7) proposed a two-step process of the kind already discussed. A further theoretical elaboration by Charlesworth and Charlesworth (1976a) showed that selection on a new mimetic pattern involves a cost due to increased conspicuousness to the predator as well as a benefit due to a higher chance of being mistaken for the model once seen. Small mutational changes are unlikely to confer a resemblance to the model sufficient to overcome the disadvantage of increased conspicuousness, and so a relatively major change in pattern is required. Subsequent to the incorporation of such a change into the population, further, more minor, changes improving the mimetic resemblance will be advantageous.

Thus, the initial stage of evolution of Batesian mimicry will almost certainly be a gene with a relatively major effect on the pattern phenotype, although a very strong degree of resemblance to the model may not be needed for protection (Brower et al. 1971). This will result in a polymorphism at the locus in question. Modifier alleles that affect the pattern of both mimetic and nonmimetic individuals (nonspecific modifiers, in the terminology of Charlesworth and Charlesworth 1976b) will be unable to spread unless they are either so closely linked to the primary locus that a strong association between the mimicry allele and modifier allele can be maintained by selection in the face of recombination, or if the selective advantage of mimicry is so large that both the mimicry and modifier alleles go to fixation (fig. 4A). The latter will be the case if the models are either highly distasteful, or very abundant relative to the mimetic species. In the former case, a

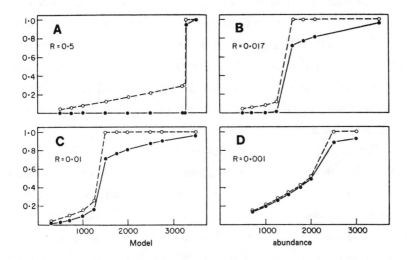

Figure 4. Changes in frequency of a gene for poor mimetic resemblance (open circles) and a nonspecific modifier (closed circles) for four different values of the recombination frequency between them (R). The dependence of the behavior of the system on the relative abundance of the model species, which controls the advantage of the mimetic resemblance, is displayed. Modified from Charlesworth and Charlesworth 1976b, fig. 1.

two-locus polymorphism with strong associations between the two loci will be maintained at low model abundance, in such a way that the population consists largely of good mimics and nonmimetic individuals. At high model abundance, both the mimicry gene and the modifier go to fixation. With intermediate ranges of model/mimic abundances, a third class of imperfect mimic, equivalent to the original imperfect mimetic form, can coexist with the other types (fig. 4B-D). There will also be selection for modifiers of genetic recombination that reduce the frequency of crossing over between the loci concerned, although this pressure is weak and the process of adjustment will be very slow compared with the processes just discussed (Charlesworth and Charlesworth 1976b).

The net result of selection in situations where polymorphism is maintained is to produce a supergene consisting of several tightly linked loci affecting different aspects of the pattern, and maintained by selection in favor

of the maximum degree of mimetic resemblance to the various models con-
cerned (Sheppard 1959, 1962; Charlesworth and Charlesworth 1976a, 1976b;
Turner 1977, 1984). This prediction is confirmed by the genetic studies of
Clarke and Sheppard (1960, 1971, 1972) on *Papilio dardanus, P. memnon* and
*P. polytes.* In the case of *P. memnon,* five closely linked genes, each con-
trolling separate components of the mimetic patterns, appear to be involved
(Clarke and Sheppard 1971).

Furthermore, the predictions of the above theory concerning the kinds
of populations that are to be expected under different relative abundance of
models and mimics are consistent with the evidence from *P. dardanus,* which
is widely distributed throughout Africa (Charlesworth and Charlesworth
1976b). The nonmimetic forms of this species are tailed, whereas the models are tailless species belonging to different, unrelated genera. Populations
polymorphic for mimetic and nonmimetic forms are found only in Ethiopia,
where the mimics are tailed. Elsewhere in Africa, all forms are tailless mimics. Taillessness is controlled by a single gene that is unlinked to the supergene controlling the wing color and pattern. Since Ethiopia is at the extreme
northern range of the species and its models, it would appear that mimetic
resemblance has a sufficiently low advantage that the allele for taillessness,
acting as a nonspecific modifier of the wing pattern component of mimicry,
has been unable to invade and cause fixation of the pattern gene, in contrast
to the situation elsewhere in Africa. In accordance with this idea, the next
most northern population (in Kenya) contains tailed mimics which exhibit
imperfectly mimetic wing patterns that are inherited as alleles at the supergene locus. This is exactly as predicted by the theory, for the case of an intermediate degree of model abundance. Elsewhere in Africa, the butterflies
are perfect mimics, as expected if model abundance is high.

In addition to nonspecific modifiers, which remain polymorphic if
closely linked to the primary locus controlling mimetic resemblance, there
may be specific modifiers that enhance the resemblance of the mimics but
have no effect on the nonmimetic genotypes. These are unconditionally ad-

vantageous, and would be expected to go to fixation (Charlesworth and Charlesworth 1976b). There is no selection for close linkage to the primary locus. The genetic studies of Clarke and Sheppard (1960) have provided abundant evidence for the existence of this kind of modifier. In particular, crosses between African butterflies and the exclusively nonmimetic *P. dardanus* found on Madagascar lead to a considerable breakdown in mimetic resemblance, and a blurring of the sharp segregation found within African populations. This demonstrates that the apparently clear-cut differences between morphs are the product of a multistep process of adjustment, in which the effects of the relatively major genes comprising the supergene have been enhanced by the fixation of numerous genes of minor effect.

**Species Differences.** Genetic studies of the basis for morphological differences and reproductive isolation between pairs of hybridizable species have been most intensively carried out in the genus *Drosophila* (Dobzhansky 1951; Coyne 1983, 1985; Val 1977; Templeton 1977, 1981). Unfortunately, morphological differences between most crossable species of this group are usually relatively slight, and are often confined to male characteristics. The detailed genetic studies of members of the melanogaster subgroup by Coyne (1983, 1985) and Coyne and Kreitman (1986), using marked chromosomes to detect linkage of genes affecting morphology to individual chromosomes or chromosome arms, indicate that there is a polygenic basis for differences in all the characters studied. The most spectacular difference between a pair of *Drosophila* species is that for head width between the Hawaiian species *D. heteroneura* and *silvestris,* the former having a hammer-shaped head which is much more pronounced in males than females. Yet the genetic distance between these species at enzyme loci is very low, indicating that they diverged relatively recently, and they are readily crossable in the laboratory. Genetic analysis of the basis for this difference is hampered by the absence of markers, but a quantitative analysis of $F_1$, $F_2$ and backcross data indicates that several autosomal genes of minor effect and an X-linked factor or factors with a more major effect are involved (Val 1977; Templeton 1977; Lande 1981).

Genetic studies of morphological differences in other species of animals and plants have been reviewed by Maynard Smith (1983), who concluded that

> the genetic basis of species differences is similar in kind to that of variation within species. Some species differences show polygenic inheritance, and others are caused by one or a few major genes. The fact that a trait behaves in a cross as if caused by a single gene does not prove that it arose in evolution by a single mutational step, because we may be observing the segregation of a gene regulating the activity of many others.

There is thus no evidence from this source that strongly favors the class 2 explanation, and many cases in which the polygenic basis for differences in morphology definitely rules it out. Many of the cases where simple genetic differences are found involve external pattern and pigmentation characteristics, where harmful pleiotropic effects on fitness are less likely than for morphological differences requiring more profound developmental changes (Charlesworth et al. 1982). However, Gottlieb (1984) has claimed that harmful pleiotropic effects of genes with major effects on morphology are less likely in plants, with their open system of development, than in animals, and cited examples of involvement of such major gene effects in variation between and within species. This claim has been severely criticized by Coyne and Lande (1985), who point out that many of Gottlieb's examples derive from artificially selected varieties, and that the evidence for single-gene control of several of his examples of natural variation is inconclusive.

There is also indirect evidence from the study of postzygotic reproductive isolation that supports the class 1 explanation of the evolution of the physiological and developmental differences between the species concerned that results in inviability or infertility of hybrids. This is the frequent strong involvement of the sex chromosomes in hybrid fitness loss (Charlesworth et al. 1987), and the concomitant tendency for the heterogametic sex to suffer disproportionately in species crosses (Haldane 1922). These observations suggest that the rate of evolution may be higher for the sex chromosomes than for autosomal regions of comparable size, producing a greater accumulation of genetic differences at sex-linked than autosomal loci. Theoretical analysis of the conditions under which the rate of gene substitution is higher for sex-linked than autosomal loci reveal that the fixation of favorable recessive or partially recessive mutations is the most likely explanation for these findings, the hybrid fitness loss being due to the fact that genes substituted independently in separate populations are not under selection to interact favorably (Charlesworth et al. 1987).

Although closely related species frequently differ with respect to gross chromosomal rearrangements (White 1978), there is little evidence that these have any causal relationship with morphological evolution, as suggested by Wilson and coworkers on the basis of a general correlation across taxa between rates of speciation and morphological evolution and the rate of chromosomal evolution (Wilson et al. 1975; Bush et al. 1977). In the first place, there is extensive chromosomal variation within species in taxa such as *Drosophila* without any noticeable morphological differences associated with this variation (Dobzhansky 1951). Secondly, morphological differences between species are often unaccompanied by chromosome rearrangements, as in the numerous homosequential species pairs in the Hawaiian *Drosophila* fauna, such as *D. heteroneura* and *D. silvestris* (Carson 1970). Third, the rate of substitution of chromosome rearrangements is very low (of the order of 1 per million years), even in groups that are evolving relatively fast at this level (Lande 1979). Unless multiple genetic effects of a kind postulated by Goldschmidt (1940) for his systemic mutations, but never observed in the laboratory, arise from the consequences of rearrangements for gene expression, it seems inconceivable that, for example, the manifold differences between humans and chimpanzees could be due simply to the dozen or so chromosome rearrangements that distinguish them (Yunis and Prakash 1982). Finally, as pointed out by Charlesworth et al. (1982) and Larson et al. (1984), there are other possible explanations for the correlation between morphological and chromosomal changes.


## Conclusions


The results of theoretical population genetics and of the experimental genetic analysis of both intra- and interspecific differences are inconsistent with the idea that class 2 processes ever play an important role in the evolution of complex adaptations. This is not to deny the possibility that mutations of relatively large effect may be involved in adaptive evolution; the examples of mimicry in butterflies (Turner 1977, 1984), and of breeding systems such as heterostyly in plants (Barrett 1989), show that this does indeed happen. Rather, the possibility that a well-integrated complex of characteristics can arise by a single mutational change seems to be ruled out. The substitution of major gene changes affecting part of a character complex is a class 1, not a class 2, process, and the examples just quoted fall into this category.

It is fair to say that the jury is still out on the question of class 1 versus class 3 explanations. There is no doubt that the Darwinian process of fixation of individually favorable traits is a process of fundamental importance in evolution, and the operation of natural selection on a variety of traits has been thoroughly documented (Endler 1986). As we have seen, the genetical theory of class 1 processes makes detailed predictions about specific systems, such as mimicry, that are validated empirically. No such validation has so far proved possible for type 3 processes. Since their operation depends critically on the magnitudes of parameters of population structure, such as local population size and rates of migration (Rouhani and Barton 1987b), these processes are likely to operate far less widely than class 1 processes. Experimental studies of subdivided populations of *Tribolium* indicate that genetic differentiation with respect to nonadditively inherited components of fitness can be produced by genetic drift in small populations even in the face of substantial migration, as postulated in the shifting-balance theory (Wade and McCauley 1984), but this does not establish that it operates under natural conditions. As discussed by Barton and Charlesworth (1984), the existence of nonadditive effects on fitness in itself does not provide a sufficient condition for evolution by peak shifts. There are, however, reasons for believing that evolution at the level of chromosome structure may be at least in part a class 3 process (Wright 1941; Lande 1979, 1985a; Walsh 1982), although class 1 processes are clearly involved in many cases, such as the inversion polymorphisms of *Drosophila* species (Dobzhansky 1951). But there is no evidence that compels us to accept a class 3 over a class 1 explanation for any evolutionary process involving morphological change. It has been suggested that punctuational patterns of change observed in the fossil record can be explained by stochastic peak shifts, since the time occupied by the transition is very short compared with the waiting time between transitions (Newman et al. 1985; Lande 1985b, 1986). Such patterns are, however, equally compatible with evolution due to selective responses of large populations to episodic environmental changes (Charlesworth et al. 1982). As far as adaptive evolution is concerned, stochastic shifts between selective peaks remain entirely hypothetical.

The empirical evidence discussed so far comes mainly from studies of intraspecific variation and from the genetic analysis of differences between crossable taxa. Advocates of the view that there is a qualitative difference between the processes of microevolution and macroevolution can always be safe in stating that genetics is powerless to shed light on the evolution of differences between higher taxa. The question of the nature of such differences has been extensively discussed in some of the classics of the "Modern Synthesis" (Simpson 1944, 1953; Rensch 1959). I can add nothing new to their arguments, which lead to the conclusion that higher taxa differ in much the same ways as lower taxa, but more profoundly and in more ways. Many examples of series of intermediate stages in the degree of complexity of adaptations such as eyes (Salwini-Plawen and Mayr 1977) are known in living forms, and examples from the fossil record of transitional forms linking major groups such as mammals and reptiles (Kemp 1982) support the notion of a Darwinian, stepwise process of evolution. Lande (1978) has made the telling argument that a saltatory mode of evolution is most likely for the loss of characteristics, such as limb reduction and loss in burrowing lizards, yet there is abundant evidence that this has proceeded by a series of gradual steps.

As Darwin wrote to Lyell 128 years ago:

> For the life of me, I cannot see any difficulty in natural selection producing the most exquisite structure, if such structure can be arrived at by gradation, and I know from experience how hard it is to name any structure towards which at least some gradations are not known. (Darwin 1958:247)

## References

Alberch, P. 1980. Ontogenesis and morphological diversification. *American Zoologist* 20: 653-67.

Barrett, S. C. H. 1989. Mating system evolution and speciation in heterostylous plants, pp. 257-83. In *Speciation and its Consequences*, ed. D. Otte and J. A. Endler. Sunderland, MA: Sinauer Associates, Inc.

Barton, N. H., and B. Charlesworth. 1984. Genetic revolutions, founder events and speciation. *Annual Review of Ecology and Systematics* 15: 133-64.

Bishop, J. A., and L. M. Cook. 1981. *Genetic Consequences of Man-made Change*. Orlando, FL: Academic Press.

Brower, L. P., J. Alcock, and J. V. Z. Brower. 1971. Avian feeding behavior and the selective advantage of incipient mimicry, pp. 261-74. In *Ecological Genetics and Evolution*, ed. E. R. Creed. Oxford: Blackwell.

Bush, G. L., S. M. Case, A. C. Wilson, and J. Patton. 1977. Rapid speciation and chromosomal evolution in mammals. *Proceedings of the National Academy of Sciences* (USA) 74: 3942-46.

Carson, H. L. 1970. Chromosome tracers of the origin of species. *Science* 168: 1414-18.

Carson, H. L. 1975. The genetics of speciation at the diploid level. *American Naturalist* 109: 73-92.

Carson, H. L. 1982. Speciation as a major reorganization of polygenic balances, pp. 411-33. In *Mechanisms of Speciation*, ed. C. Barigozzi. New York: Liss.

Carson, H. L., and A. R. Templeton. 1984. Genetic revolutions in relation to speciation phenomena: The founding of new populations. *Annual Review of Ecology and Systematics* 15: 97-131.

Charlesworth, B., J. A. Coyne, and N. H. Barton. 1987. The relative rates of evolution of sex chromosomes and autosomes. *American Naturalist* 130: 113-46.

Charlesworth, B., R. Lande, and M. Slatkin. 1982. A neo-Darwinian commentary on macro-evolution. *Evolution* 36: 474-98.

Charlesworth, B., and S. Rouhani. 1988. The probability of peak shifts in a founder population. II. An additive polygenic trait. *Evolution* 42: 1129-45.

Charlesworth, D., and B. Charlesworth. 1976a. Theoretical genetics of Batesian mimicry. I. Single-locus models. *Journal of Theoretical Biology* 55: 283-303.

Charlesworth, D., and B. Charlesworth. 1976b. Theoretical genetics of Batesian mimicry. II. Evolution of supergenes. *Journal of Theoretical Biology* 55: 305-24.

Clarke, C. A., and P. M. Sheppard. 1960. The evolution of mimicry in the butterfly *Papilio dardanus*. *Heredity* 14: 163-73.

Clarke, C. A., and P. M. Sheppard. 1971. Further studies on the genetics of the mimetic butterfly *Papilio memnon* L. *Philosophical Transactions of the Royal Society of London*. *Series B*. 263: 35-70.

Clarke, C. A., and P. M. Sheppard. 1972. The genetics of the mimetic butterfly *Papilio polytes* L. *Philosophical Transactions of the Royal Society of London*. *Series B*. 263: 431-58.

Coyne, J. A. 1983. Genetic basis of difference in genital morphology among three sibling species of the genus *Drosophila*. *Evolution* 37: 1101-18.

Coyne, J. A. 1985. Genetic studies of three sibling species of *Drosophila* with relationship to theories of speciation. *Genetical Research* 46: 169-92.

Coyne, J. A., and M. Kreitman. 1986. Evolutionary genetics of two sibling species of *Drosophila*, *D. simulans* and *D. mauritiana*. *Evolution* 40: 673-91.

Coyne, J. A., and R. Lande. 1985. The genetic basis of species differences in plants. *American Naturalist* 126: 141-45.

Darwin, C. R. 1865. On the sexual relations of the three forms of *Lythrum salicaria*. *Journal of the Proceedings of the Linnean Society*. *Botany* 8: 169-96.

Darwin, F. 1958. *The Autobiography of Charles Darwin and Selected Letters*. New York: Dover.

Dobzhansky, T. 1951. *Genetics and the Origin of Species*. 3d ed. New York: Columbia University Press.

Endler, J. A. 1986. *Natural Selection in the Wild*. Princeton: Princeton University Press.

Fisher, R. A. 1930. *The Genetical Theory of Natural Selection.* Oxford: Oxford University Press.

Goldschmidt, R. B. 1940. *The Material Basis of Evolution.* New Haven: Yale University Press.

Goldschmidt, R. B. 1945. Mimetic polymorphism, a controversial chapter of Darwinism. *Quarterly Review of Biology* 20: 147-64, 205-30.

Gottlieb, L. D. 1984. Genetics and morphological evolution in plants. *American Naturalist* 123: 681-709.

Gould, S. J. 1980. Is a new and general theory of evolution emerging? *Paleobiology* 3: 115-51.

Gould, S. J. 1982. Darwinism and the expansion of evolutionary theory. *Science* 216: 380-87.

Hadorn, E. 1961. *Developmental Genetics and Lethal Factors.* London: Methuen.

Haldane, J. B. S. 1922. Sex-ratio and unisexual sterility in hybrid animals. *Journal of Genetics* 12: 101-9.

Haldane, J. B. S. 1924. A mathematical theory of natural and artificial selection. Part I. *Transactions of the Cambridge Philosophical Society* 23: 19-41.

Haldane, J. B. S. 1927. A mathematical theory of natural and artificial selection. Part V. Selection and mutation. *Proceedings of the Cambridge Philosophical Society* 28: 838-44.

Haldane, J. B. S. 1932. *The Causes of Evolution.* London: Longmans Green.

Kemp, T. S. 1982. *Mammal-like Reptiles and the Origin of Mammals.* New York: Academic Press.

Lande, R. 1978. Evolutionary mechanisms of limb loss in tetrapods. *Evolution* 32: 73-92.

Lande, R. 1979. Effective deme sizes during long-term evolution estimated from rates of chromosomal evolution. *Evolution* 33: 234-51.

Lande, R. 1981. The minimum number of genes contributing to quantitative variation between and within populations. *Genetics* 99: 541-53.

Lande, R. 1985a. The fixation of chromosome rearrangements in a subdivided population with local extinction and colonization. *Heredity* 54: 323-32.

Lande, R. 1985b. Expected time for random genetic drift of a population between stable phenotypic states. *Proceedings of the National Academy of Sciences* (USA) 82: 7641-45.

Lande, R. 1986. The dynamics of peak shifts and the pattern of morphological evolution. *Paleobiology* 12: 343-54.

Larson, A., E. M. Prager and A. C. Wilson. 1984. Chromosomal variation, speciation and morphological change in vertebrates: The role of social behavior. *Chromosomes Today* 8: 215-28.

Løvtrup, S. 1974. *Epigenetics.* New York: John Wiley.

Maynard Smith, J. 1975. *The Theory of Evolution.* 3d ed. London: Penguin Books.

Maynard Smith, J. 1983. The genetics of stasis and punctuation. *Annual Review of Genetics* 17: 11-25.

Mayr, E. 1954. Change of genetic environment and speciation, pp. 157-80. In *Evolution as a Process,* ed. J. Huxley, A. C. Hardy, and E. B. Ford. London: Allen & Unwin.

Mayr, E. 1963. *Animal Species and Evolution.* Cambridge: Harvard University Press.

Newman, C. M., J. E. Cohen and C. Kipnis. 1985. Neo-Darwinian evolution implies punctuated equilibria. *Nature* 315: 400-401.

Nicholson, A. J. 1927. A new theory of mimicry in insects. *Australian Zoologist* 5: 10-104.

Provine, W. B. 1971. *The Origins of Theoretical Population Genetics.* Chicago: The University of Chicago Press.

Punnett, R. C. 1915. *Mimicry in Butterflies.* Cambridge: Cambridge University Press.

Rensch, B. 1959. *Evolution Above the Species Level.* New York: Columbia University Press.

Rouhani, S., and N. H. Barton. 1987a. The probability of peak shifts in a founder population. *Journal of Theoretical Biology* 126: 51-62.

Rouhani, S., and N. H. Barton. 1987b. Speciation and the "shifting-balance" in a continuous population. *Theoretical Population Biology* 31: 465-92.

Salwini-Plawen, L. V., and E. Mayr. 1977. The evolution of photoreceptors and eyes. *Evolutionary Biology* 10: 207-63.

Sheppard, P. M. 1959. The evolution of mimicry: A problem in ecology and genetics. *Cold Spring Harbor Symposium on Quantitative Biology* 24: 131-40.

Sheppard, P. M. 1962. Some aspects of the geography, genetics and taxonomy of a butterfly, pp. 135-52. In *Taxonomy and Geography,* ed. D. Nichols. London: Systematics Association.

Sheppard, P. M., J. R. G. Turner, K. S. Brown, W. W. Benson, and M. C. Singer. 1985. Genetics and the evolution of Müllerian mimicry in *Heliconius* butterflies. *Philosophical Transactions of the Royal Society of London. Series B.* 308: 433-613.

Simpson, G. G. 1944. *Tempo and Mode in Evolution.* New York: Columbia University Press.

Simpson, G. G. 1953. *The Major Features of Evolution.* New York: Columbia University Press.

Stanley, S. M. 1979. *Macroevolution: Pattern and Process.* San Francisco: W. H. Freeman.

Templeton, A. R. 1977. Analysis of head shape differences between two interfertile species of Hawaiian *Drosophila. Evolution* 31: 630-41.

Templeton, A. R. 1980. The theory of speciation by the founder principle. *Genetics* 92: 1011-38.

Templeton, A. R. 1981. Mechanisms of speciation – A population genetic approach. *Annual Review of Ecology and Systematics* 12: 23-48.

Thomson, K. S. 1987. History, development and the vertebrate limb. *American Scientist* 75: 518-20.

Turner, J. R. G. 1977. Butterfly mimicry: The genetical evolution of an adaptation. *Evolutionary Biology* 10: 163-206.

Turner, J. R. G. 1984. Darwin's coffin and Doctor Pangloss – Do adaptationist models explain mimicry? pp. 313-61. In *Evolutionary Ecology*, ed. B. Shorrocks. Oxford: Blackwell.

Val, F. C. 1977. Genetic analysis of the morphological differences between two interfertile species of Hawaiian *Drosophila*. *Evolution* 31: 611-29.

Wade, M. J., and D. E. McCauley. 1984. Group selection: The interaction of local deme size and migration in the differentiation of small populations. *Evolution* 38: 1047-58.

Walsh, J. B. 1982. Rate of accumulation of reproductive isolation by chromosomal rearrangements. *American Naturalist* 120: 510-32.

White, M. J. D. 1978. *Modes of Speciation*. San Francisco: W. H. Freeman.

Wilson, A. C., G. L. Bush, S. M. Case, and M. C. King. 1975. Social structuring of mammalian populations and rate of chromosomal evolution. *Proceedings of the National Academy of Sciences* (USA) 72: 5061-65.

Wright, S. 1931. Evolution in Mendelian populations. *Genetics* 16: 97-159.

Wright, S. 1932. The roles of mutation, inbreeding, crossbreeding and selection in evolution. *Proceedings of the 6th International Congress of Genetics* 1: 356-66.

Wright, S. 1941. On the probability of fixation of reciprocal translocations. *American Naturalist* 75: 513-22.

Wright, S. 1949. Adaptation and selection, pp. 365-89. In *Genetics, Paleontology and Evolution*, ed. G. L. Jepsen, E. Mayr, and G. G. Simpson. Princeton: Princeton University Press.

Wright, S. 1968. *Evolution and the Genetics of Populations. I. Genetic and Biometric Foundations*. Chicago: The University of Chicago Press.

Wright, S. 1980. Genic and organismal selection. *Evolution* 34: 825-43.

Yunis, J. J., and O. Prakash. 1982. The origin of man: A chromosomal pictorial legacy. *Science* 215: 1525-30.

# Heterochrony and Other Mechanisms of Radical Evolutionary Change in Early Development

*Rudolf A. Raff, Brian A. Parr, Annette L. Parks, and Gregory A. Wray*

That morphological evolution requires changes in the programs of development underlying morphogenesis is by now a well-recognized truism as well as a subject of growing popularity among evolutionary biologists. There have been two main lines of approach, both combining comparative with experimental methodologies. The first approach, based on theoretical considerations, presupposes that organisms cannot respond entirely freely to selection, but that the responses are limited or channeled by the nature of genetic and developmental systems. These are "developmental constraints" to evolution (Alberch 1982; Maynard Smith et al. 1985). The hypothesis of developmental constraints is of obvious appeal because it provides some directional rules to govern evolution. These may underlie long-term evolutionary trends, and they should allow predictions about possible courses of evolution open to particular lineages. Furthermore, the concept of developmental constraints has served as a basis for empirical research on the evolution of developmental processes (Alberch 1987).

The second approach is more general. Changes in developmental programs are sought for particular lineages being compared, and the mechanisms by which developmental programs have come to differ are defined. Definition is possible at several hierarchical levels as we seek to link morphological change with the genetic changes that ultimately underlie them. This approach is grounded in the concept of dissociability. That is, developmental processes which happen together or sequentially in time are not necessarily tightly coupled mechanistically and may be shifted relative to each other in evolution without disrupting development (Needham 1933). Heterochrony is the most familiar kind of dissociation in which relative timing of two developmental processes undergoes an evolutionary shift.

A few appropriate experimental systems are now being examined in various laboratories (see Raff and Raff 1987), and certain requirements for research on the evolution of developmental systems can be listed (Raff 1988): (1) Although valuable information on development of extinct organisms has been drawn from the fossil record, insufficient evidence is available for detailed comparison with ontogenies of living descendants. Thus, in general, comparisons must be made between related living organisms. (2) The organisms compared should be as closely related as possible to avoid irrelevant noise. (3) The features or processes being compared must be demonstrably homologous. (4) Developmental processes must be accessible to experimentation and preferably sufficiently simple that the investigator is not overwhelmed by complex interactions. (5) The polarity of evolutionary change must be known.

Echinoids (sea urchins and sand dollars) meet these criteria, and provide an excellent system for studying the evolution of development. The approximately 1000 living species of sea urchins exhibit a spectacular evolutionary diversity in modes of early development. In the typical pattern of sea-urchin development, a complex planktotrophic feeding larva, the pluteus, is produced as a result of the early developmental events of cleavage, blastulation, gastrulation, and larval morphogenesis. The juvenile rudiment, which ultimately gives rise to the adult, forms from mesodermal and ectodermal structures on the left side of the larva (fig. 1) (Okazaki 1975). However, an estimated 20% of sea urchin species have evolved some mode of direct development (Raff 1987). Direct development has evolved independently in several lineages, but all direct-developing echinoids exhibit large yolky eggs, loss of larval feeding, and modifications or loss of typical pluteus features. In some instances novel larval features have evolved. In other, more extreme cases, all larval features are lost, and development involves only the production of adult sea urchin features.

Loss of the feeding-pluteus larva is strongly correlated with egg size. Typical feeding plutei are produced by species with eggs 60-200 μm in diameter. Of the approximately 100 species for which information on development is available (Emlet et al. 1987), about eighty develop through a pluteus. Only two species that produce a feeding pluteus have eggs with a diameter exceeding 200 μm. One of these, with an egg diameter of 280 μm, is a facultative feeder; it has sufficient yolk reserves to metamorphose even if starved (Emlet 1986). Species with larger eggs, about 300-350 μm in diameter, produce abbreviated plutei with larval skeletons, but do not develop functional guts, and exhibit rapid development of the juvenile sea urchin rudiment. The largest eggs, 400-2000 μm in diameter, either produce nonfeeding floating larvae or are brooded by the mother and develop directly into juvenile adults without development of a skeleton or other features of the pluteus. These modes of development are diagrammed in figure 1.

## Dissociation in Evolution of Early Development

The prevalence of direct development in various groups of sea urchins suggests that early development in sea urchins occurs by highly dissociable processes. Obviously, although such a dissociation is not disruptive, it can yield an altered ontogeny, with resulting changes in relative timing of events or alterations in location of features in the embryo.

Two kinds of dissociation may have occurred in the evolution of direct development. The first is dissociation of gene expression patterns. Embryonic development can be envisaged as consisting of two genetic compartments. One compartment consists of a gene expression program for the development of the pluteus larva; the other consists of a gene expression program for the juvenile adult. If these two gene expression programs are loosely linked, the development of a juvenile may not require prior expres-

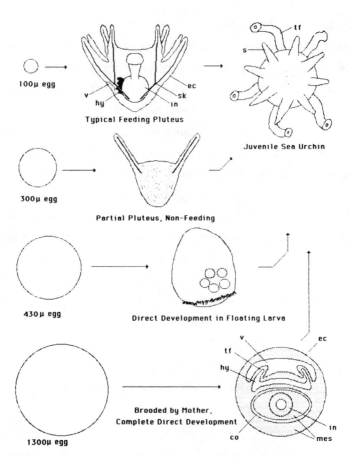

Figure 1. Typical and direct patterns of development in echinoids. Abbreviations: co, coelom; ec, ectoderm; hy, hydrocoel; in, intestine; mes, mesenchyme; sk, skeleton; tf, tube foot; v, vestibule.

sion of the larval program. Since gene expression in the juvenile rudiment is very poorly known, this remains a speculative hypothesis. However, it is clear that there are genes in sea urchins which are expressed only in early development (Davidson 1986; Hursh et al. 1987).

The second type of dissociation that may have occurred during the evolution of direct development is dissociation of cell-cell interactions. Regardless of what gene expression patterns may have changed, it is clear that the behaviors of various groups of cells, or cell lineages, are greatly modified with respect to each other in direct-developing species (Raff 1987). This dissociation between differentiating cell lineages requires that, as cell lineages differentiate in early development, they do so relatively autonomously. As the patterns of cellular behavior or interaction change in the evolution of direct development, pluteus structures are deleted and replaced by new larval features. Consequently, some larval cell types differentiate incompletely in direct-developing embryos. For example, endodermal cells do not form a functional gut. Other cell types follow a distinctly different course of differentiation. Thus, primary mesenchyme cells go directly to the adult program of skeletogenesis (Williams and Anderson 1975; Raff 1987; Parks et al. 1988).

The concept of developmental constraint suggests that adaptations produced in response to selection may be limited by the properties of the ancestral developmental program (Alberch 1982, 1985; Gould 1982; Maynard Smith et al. 1985). Some hypothetically adaptive responses to selection may not be attainable because the starting genetic control networks cannot reach certain states, interactions between developmental processes may be too complex to modify, or programs follow epigenetic rules that exclude certain results. Attainment of other descendant states may be favored by features of the system that bias the response to selection in particular directions. Descendant states which are both adaptive and attainable by some evolutionary trajectory that does not violate developmental constraints will contain the morphologies actually produced.

It has been suggested that evolutionary changes will occur more readily in later stages of development because any change in early development would affect all subsequent ontogeny (Gould 1977; Raff and Kaufman 1983; Buss 1987). This hypothesis predicts that early stages will be particularly conservative in evolution. However, empirical evidence does not bear this out (Sedgwick 1894; Garstang 1928; Roth 1984; Elinson 1987; Raff 1987). In fact, in some organisms such as vertebrates, early development may actually

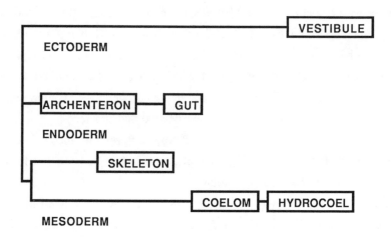

Figure 2. Timing of development of larval features in different cell lineages of typically developing echinoids.

be less integrated and constrained than some later stages (Elinson 1987). This also may be the case in sea urchins, as illustrated by the cellular events of early echinoid development. The three classic "germ layers" roughly define three major groups of cell lineages as they appear at the fourth cleavage of typical sea urchin embryos (fig. 2).

At the fifth cleavage, two distinct lineages of mesodermal cells arise, primary mesenchyme and coelomic precursors. The diagram is an oversimplification because other cell lineages arise in these embryos as well (Davidson 1986). However, these other lineages do not affect the argument based on the timing of appearance of the larval features considered in the diagram. In typical development the order of events is: Archenteron - Skeleton - Gut - Coelom - Hydrocoel - Vestibule. This definite order is consistent with two interpretations. The first possibility is that each lineage differentiates entirely separately from each other, and the timing of events is entirely autonomous within individual cell lineages. The second possibility is that the events occurring in each lineage directly influence other lineages by inductive interactions to produce a tight integration of timing. The real situation probably resides somewhere between these extremes. Autonomous differentiation has been demonstrated for at least one cell lineage, the primary-mesenchyme (Okazaki 1975). There also is evidence for inductive interactions in sea urchins (Horstadius 1973), but it is unclear to what extent interactions govern the events diagrammed in figure 2. For example, the development of the hydrocoel appears to be a direct consequence of formation of the coelom. Conversely, appearance of the vestibule is apparently not dependent upon prior formation of the hydrocoel (Czihak 1960, 1965). The shifts in relative timing of these events observed with direct-developing sea urchins (see below) suggests that inductive interactions are not absolutely required, and that cell lineage behavior is evolutionarily dissociable in early echinoid development.

### Heterochrony Model for Evolution of Direct Development

Although detailed descriptions of direct-developing embryos are still few, there is sufficient information on a few species to permit a comparison of the relative order of developmental events with those of typical species (Raff 1987). These comparisons reveal dramatic heterochronies. There are condensations of the appearance of typical features of the pluteus. These condensations are variable: the gut is eliminated in all species examined, whereas a partial pluteus skeleton forms in some and is completely eliminated in others (Raff 1987). Further, there is an early onset of appearance and accelerated development of adult features relative to gastrulation and development of retained larval features. Thus, some features that typically appear only in a fully differentiated pluteus days after gastrulation may appear as early as late gastrulation in direct development.

These obvious heterochronies provide the foundation for a simple heterochrony model as a working hypothesis for the evolution of direct development. This model suggests that heterochronies will be manifested at the level of individual cell lineages. These heterochronies result in a novel larval morphology in which some pluteus features are eliminated and some adult features make an early appearance. As will be discussed below, there is no reason to suppose that all cell lineages undergo the same heterochronies. Further, we expect that individual cell lineages will undergo changes in gene expression consistent with the heterochronies they express in behavior and morphogenesis. The relative timing of developmental events for a typical developing species and two direct developers illustrates these points.

1. Typical
Archenteron - Skeleton - Gut - Coelom - Hydrocoel - Vestibule

2. *Heliocidaris erythrogramma*
Archenteron - Coelom - Hydrocoel - Vestibule

3. *Peronella japonica*
Archenteron - Skeleton - Vestibule - Coelom - Hydrocoel

The development of *H. erythrogramma* illustrates the condensation of larval features, early onset and acceleration of adult features as described above. Comparison of the development of *P. japonica* with *H. erythrogramma* shows two things. First, heterochronies can occur differentially between cell lineages. Thus, larval gut and skeleton, which derive from different lineages, are not linked to each other. Second, other more complex heterochronies can also occur, in this case involving an early onset of vestibule formation relative to the appearance of coelom and hydrocoel.

## Defining Homologous Cell Lineages

Heterochrony is generally discussed in systems terms. The classic heterochronies of de Beer (1958) and Gould (1977) consist of statements on the relative timing of gonadal versus somatic maturation; homologies are implicit and details of mechanism are ignored. This approach has successfully defined a set of phenomena so that they become recognizable and coherent.

Yet, heterochronies do not occur *in vacuo.* The mechanisms which yield the changes ultimately recognized as heterochronous results occur in the developmental programs of actual cells within the embryo. It is particularly crucial to recognize this point in dealing with heterochronies in early development, because the critical steps of early development involve the specification and differentiation of discrete cell lineages within the embryo (Stent 1985). Evolutionary modification of early development requires changes in the specification, location, behavior, or fate of particular cell lineages (Raff 1987).

All comparative analyses of evolutionary relationships require that homologies be identified. This is not always a straightforward matter (Roth 1988), and it is particularly difficult with larval, and more especially, embryonic features. There is some controversy over what criteria should be used to identify homologous structures (Remane 1971; Bock 1974; Riedel 1978; Roth 1984). Three generally used criteria are: position in a comparable system of features, similarity of structure and organization of a feature, and identification of transitional forms (Remane 1971; Riedel 1978). For example, the overall successful application of such criteria as comparative anatomy, position in the body, and a fossil record leave little doubt about homology of vertebrate forelimbs. The dangers of using developmental features in establishing homology have been pointed out by Wilson (1894) and de Beer (1971) because some clearly homologous adult structures can be shown to arise from different embryonic regions and to express apparently different modes of development. This may pose a particular risk in evaluating features of early development. However, the general criteria for evaluating homology outlined above do seem applicable if used cautiously. Homology of developmental processes is certainly visible in well-studied examples such as mammalian ear bones and vertebrate limb buds (de Beer 1971; Roth 1984). Homologies of cells in embryos of related nematodes also have been established from the strict lineage and fate patterns of these embryos (Sternberg and Horvitz 1981). Finally, homology between the genes that regulate homologous developmental processes also are beginning to emerge. The best examples are provided by the homoeotic genes that govern the development of insect segmental identity. DNA sequences in the homoeotic genes of the Antennapedia complex of *Drosophila melanogaster* are homologous with genes of similar function in other species in this genus and even with other dipterans (Seeger and Kaufman 1987). These genes also are homologous to genes with similar functions in the beetle *Tribolium* (Beeman 1987; Akam 1987).

To understand the mechanisms underlying evolutionary modifications of cell lineage behavior clearly requires unambiguous identification of homologous cell lineages in embryos with distinct styles of development. The use of closely related species for comparisons of developmental processes should maximize the probability that homologous, albeit modified, cell lineages can be identified and compared. The relatively simple histology of early stages of development further reduces the chance of confusion because of the limited number of cell types present. The simple morphology of early stages simplifies localization of cells and their sites of origin. Given realistic criteria, we should be able to identify homologous cell lineages in typical and direct-developing embryos and larvae (Raff 1987, 1988). A set of criteria are presented in table 1.

Table 1

Criteria for homology between cell lineages

---

1. Same cell precursors or embryonic regions
2. Shared features of cell organization
3. Shared patterns of cell lineage-restricted gene expression, although elements are lost, truncated, or shifted in time
4. Similar final fates

---

We are applying these criteria in comparing the developmental programs of a direct-developing sea urchin, *Heliocidaris erythrogramma*, with its typically developing congener, *H. tuberculata*, and another more distantly related typical developing species, *Strongylocentrotus purpuratus.* It should be noted that despite the root word *cidaris* in the generic name *Heliocidaris,* this genus is not a member of the primitive sea urchin family Cidaridae, but is a euechinoid. Embryology and gene expression in early development of *S. purpuratus* have been studied extensively, and cell fate mapping studies in this species have provided a fate map for the cells of the typical euechinoid sea urchin embryo (reviewed by Davidson 1986).

The thorough description of the embryology of *H. erythrogramma* by Williams and Anderson (1975) provides a basis for provisional recognition of several cell lineages in this embryo. Despite the differences in larval development exhibited by this species, the basic groups of cells defined in typical sea urchins also exist; they arise from the same general locations in the early embryo, and they preserve many of the same structural features, behaviors, and fates as their presumed homologs. To define these lineages in detail, we have begun a series of cell lineage tracing experiments in *H. erythrogramma* (Wray and Raff, in press).

These experiments consist of the injection of fixable fluorescent dyes of high molecular weight into individual cells of the embryos, which raises the embryo to a later stage. The embryos are then fixed, and examined by fluorescence microscopy to determine which cells contain fluorescent dye. Various questions can be addressed in this way. For example, injection of one of the two cells produced by the first cleavage of the embryo allows us to determine the relationship of the first cleavage axis to the axes of the larva. Detailed fate determinations come from injections into individual embryonic cells. Thus, injection of each of the cells of the 8-cell embryo gives a map defining four quarters in both animal and vegetal halves of the embryo. This stage is important because such a map already exists for *S. purpuratus* (Cameron et al. 1987). Injection of cells at yet more advanced stages allows finer resolution of fates.

Although the cell lineage tracing experiments are not yet complete, we have already found that some aspects of cell fate and axis determination have remained constant during the evolution of direct development in *H. erythrogramma*, whereas others have been extensively modified. The spatial relationship between embryonic and adult axes have been retained. Similarly, the fates of blastomeres in the animal half of the *H. erythrogramma* embryo closely resemble those in *S. purpuratus*. However, the fates of vegetal blastomeres show several modifications. For example, the time at which the primary mesenchyme cell lineage is clonally segregated is delayed, a heterochrony in cell fate commitment. In addition, the fates of vegetal blastomeres exhibit an early commitment to distinct dorsal and ventral fates: only ventral cells at the vegetal pole give rise to pigment cells. This break in symmetry is not observed in the typical embryo. The radical evolutionary modification in cell fates among vegetal blastomeres is not surprising since these cells give rise to the very structures that are most modified in the *H. erythrogramma* embryo.

## Molecular Heterochrony in Cell Lineages

A further approach to defining mechanisms of evolutionary changes in cell lineages is to use cell lineage-specific molecular probes. Such probes yield two kinds of information. First, probes of known cell lineage specificity in typical sea urchins can be used to help identify cellular homologs in the direct-developing species. This use of probes, such as monoclonal antibodies, is straightforward in principle, although potentially confounded if heterotopies (changes in site of expression) occur. The second use of molecular probes is to determine how gene expression has changed in conjunction with observed changes in embryology.

The questions of constraint raised earlier apply equally to modifications of gene expression in early development. Here, too, recent studies indicate that regulatory mechanisms are not necessarily tightly constrained in early echinoid development.

All echinoderms share a tandem repeat cluster of histone genes expressed in early development. These genes also are expressed in oogenesis of euechinoids to produce maternal mRNAs, but not in the cidaroids or non-echinoid echinoderms (Raff et al. 1984). Furthermore, if these maternal transcripts are experimentally removed from euechinoid eggs before the eggs are fertilized, they nevertheless undergo normal development to the pluteus stage (Wells et al. 1986). In addition, several heterochronies and heterotopies in the expression of gene products during early development have occurred in typically developing echinoids (Wray and McClay, in press b). As shown in figure 3, the protein Meso 1 has undergone both temporal and spatial alterations in expression. In the regular echinoids examined (top row, fig. 3), Meso 1 appears at the time of mesenchyme ingression and in both primary and secondary mesenchyme cells. In a sand dollar (middle row, fig. 3) the same set of cells expresses the protein, but the time of its appearance is delayed. Finally, in a cidaroid, time of appearance is delayed, and expression is confined to secondary mesenchyme (bottom row, fig. 3).

The molecular heterochronies described above do not appear to have any effect on morphogenesis or on the ultimate production of a typical pluteus. What they show is that variability in gene expression patterns are common in early typical echinoid development. Under conditions which select for evolution of a different mode of early development, such as seen in the examples of direct development, this lack of constraint in gene expression may be important in remodeling of developmental processes that require changes in timing of expression of these genes.

In fact, molecular heterochronies are evident in the evolution of direct development. We have documented one instance of a heterochronous shift in gene expression in the primary mesenchyme cell lineage of the direct-developing sea urchin *H. erythrogramma* (Parks et al. 1988). The first three cleavages following fertilization of *H. erythrogramma* resemble those of typical developers, but the fourth cleavage is equal so that no micromeres are produced. Instead of the approximately 32 primary mesenchyme cells that

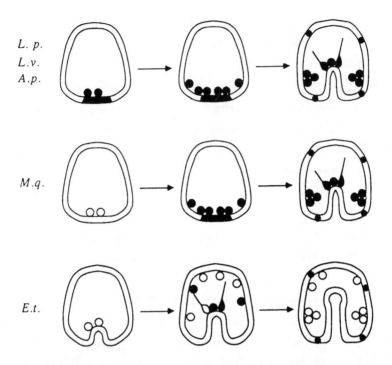

Figure 3. Temporal and spatial patterns of appearance of the mesenchymal antigen Meso 1 in five species of echinoids. Top row, regular echinoids; middle row, a sand dollar; bottom row, a cidaroid. Species are *A.p., Arbacia punctulata; E.t., Eucidaris tribuloides; L.p., Lytechinus pictus; L.v., L. variagatus; M.q., Mellita quinquesperferata.* Open circles in the diagram represent mesenchyme cells not expressing the Meso 1 antigen, whereas closed circles represent mesenchyme cells that are expressing.

ingress into the blastocoel of typically developing sea urchins, hundreds of *H. erythrogramma* mesenchyme cells enter the blastocoel prior to gastrulation. As in typical developers, the *H. erythrogramma* mesenchyme cells eventually secrete skeletal structures. However, these are adult skeletal elements that are produced much earlier than in typical developing species (2-3 days versus several weeks), and secretion of a larval skeleton never occurs. Therefore, the behavior of the mesenchyme cell lineage of *H. erythrogramma* is drastically altered by deletion of its larval morphogenetic program and an early onset of adult morphogenesis.

These changes in mesenchyme cell behavior and morphogenesis are

TYPICAL

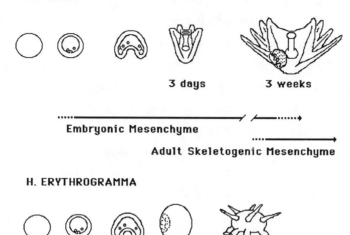

3 days          3 weeks

Embryonic Mesenchyme

Adult Skeletogenic Mesenchyme

H. ERYTHROGRAMMA

3.5 days

Adult Skeletogenic Mesenchyme

Figure 4. Programs of expression of the primary mesenchyme cell lineage-specific protein msp130 in typical sea urchins and in *H. erythrogramma*. The arrows represent the period of synthesis of msp130 in primary mesenchyme cells engaged in either formation of larval skeleton or adult skeletal elements. Synthesis is detected both by probes for msp130 mRNA and antibody staining of sections of embryos.

paralleled by alterations in gene expression in *H. erythrogramma* (fig. 4). Msp130 is a cell surface glycoprotein whose expression is restricted to the skeletogenic primary mesenchyme cell lineage of sea urchin embryos (Leaf et al. 1987; Anstrom et al. 1987; Wray and McClay, in press a). It is also present in the skeleton-secreting tissues of juveniles and adults (Parks et al. 1988). As probes for msp130 we have monoclonal antibodies to a sulfated polysaccharide modification of the protein as well as antibodies to portions of the polypeptide chain itself. In typically developing species, such as *S. purpuratus* and *H. tuberculata*, msp130 is first evident at the mesenchyme blastula stage when the primary mesenchyme cells are beginning to ingress into the blastocoel. In contrast, *H. erythrogramma* embryos do not express msp130 at detectable levels until the end of gastrulation, long after numerous mesenchyme cells have entered the blastocoel. The relative timing of msp-130 in typical sea urchins and *H. erythrogramma* is diagrammed in figure 4. Typical embryos have two temporally and spatially distinct modes of expression of msp130. *H. erythrogramma* has only one – the adult pattern. Thus, although the first appearance of msp130 appears delayed, what has in fact occurred is that the embryonic program of expression has been deleted and the adult program exhibits an early onset.

There are possible alternate explanations for this heterochrony. Msp-130 might have only a single continuous program of expression in typically developing sea urchins, and this program is initiated later in *H. erythrogramma* because the protein is not "needed" until a more advanced developmental stage. Alternatively, the appearance of msp130 may be controlled by two distinct programs (e.g., two separate enhancer/promoter elements), one responsible for larval expression, the other for adult. According to this hypothesis, not only has a larval program of msp130 expression been deleted in *H. erythrogramma*, but an adult program has been turned on earlier. At the level of resolution provided by our research to date, both hypotheses would predict parallel changes in morphogenesis of the mesenchyme cell lineage of *H. erythrogramma* and in mesenchyme lineage-specific gene expression. Further information about the regulation of msp130 will be necessary to decide this evolutionary issue.

## Nonheterochronous Mechanisms

Good ideas can sometimes come to be taken as revealed truths because of their aesthetic appeal. Such a canonization is close for the concept of heterochrony as a universal explanation for evolutionary changes in development. Because of the prevalence of heterochronies in the evolutionary literature and the ease with which heterochronies can be recognized, there is a general perception that essentially all evolutionary modifications of development involve heterochronies. This is so because heterochronous results can arise from mechanisms that primarily displace timing as well as from events that do not involve timing changes per se, but result in apparent heterochronies when viewed from the final result (Raff and Wray, submitted).

The simple heterochrony model outlined above serves to describe many of the changes seen in the evolution of direct development in echinoids, and appears to be borne out at the cellular level. However, when development is closely examined, it is clear that these heterochronous changes are not the whole story. The evolution of a large egg seems to be a necessary step in the achievement of direct development in sea urchins of all groups. Unexpectedly, the sperm heads of these forms also are unusually long (up to 5 times as long as related typical developers) (Raff et al. 1988). Important changes in early development of *H. erythrogramma* are not limited to heterochronies (Williams and Anderson 1975; Raff 1987; Parks et al. 1988; Wray and Raff, in press). They include the substitution of a symmetric fourth cleavage for the striking asymmetric fourth cleavage division of typical embryos, a modified pattern of larval cell fates from cells of the 16-cell embryo vegetal tier, the production of a wrinkled blastula, and massive segregation of yolk into the blastocoel. The effects of these nonheterochronous modifications on larval development are as profound as the heterochronous changes noted above.

Do nonheterochronous mechanisms have heterochronous results? In at least one instance the answer is yes. The cell cleavage dynamics of *H. erythrogramma* are dramatically different from those of *H. tuberculata* (Parks et al. 1988). Both species exhibit the same cell division rate during early cleavage. By the 500-cell blastula stage, *H. tuberculata* cell cleavage has slowed and the gastrula stage achieved 10 hours later has only approximately 1000 cells. This pattern of cleavage is seen in other typical echinoid embryos. But, the *H. erythrogramma* gastrula has about 14,000 cells, resulting from a longer period of active cell division. The cause is probably related to the nucleocytoplasmic ratio controls of cell division observed in other embryos (Newport and Kirschner 1982a, 1982b). This change in nucleocytoplasmic ratio resulting from a large egg cytoplasm is hardly a heterochronous mechanism, but the result is clearly heterochronous as measured in terms of cell division rate and duration relative to stage.

### Direct Development Occurs Commonly in Echinoid Evolution

We have available at least partial information on the modes of development of nearly 200 of the more than 900 known living species of sea urchins (tabulated by Emlet et al. 1987). Approximately 20% exhibit modes of development distinct from the typical feeding pluteus. There are several striking features in these data. The proportion of direct-developing (lecithotrophic and brooding) species increases in deep-water and south-polar faunas. Lecithotrophic larvae develop from large yolky eggs, and although they develop suspended in the water column, they do not feed. Brooders also develop directly from large yolky eggs, but the larvae are retained among the spines or in special brood pouches on the mother's test. Although there are no direct-developers in arctic faunas or in the shallow-water faunas of North America and Europe, direct-developers are prevalent in the western and southern Pacific. At least four direct-developers are known from Japan, and the Australian and Antarctic faunas are rich in direct-developers. The Antarctic species are brooders, whereas the majority of Australian species produce planktonic nonfeeding larvae.

Although the Australian and Antarctic sea urchin faunas are south Pacific in distribution, they are not closely related (Mortensen 1928-1951). The Australian direct-developers all belong to groups with Indo-Pacific affinities. The Antarctic groups represent distinct radiations. Australian direct-developers are mainly found among the cidaroids, and among regular sea urchins in the orders Temnopleuroida and Echinoida. The Australian brooders include one clypeasteroid (sand dollar) and one spatangoid (heart urchin).

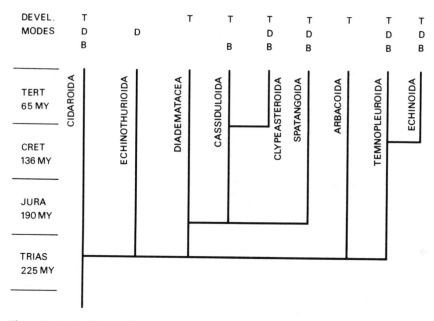

Figure 5. A simplified phylogeny of living echinoid orders. Developmental modes described in each order are given as T = typical pluteus larva; D = direct lecithotrophic larvae; B = brooded direct-developing larvae. Geologic periods indicated are Triassic, Jurassic, Cretaceous and Tertiary. The times are age in millions of years before the present.

The Antarctic brooding sea urchins are all cidaroids or heart urchins.

Information on development is available for at least some representatives of each major sea-urchin lineage shown in figure 5. Direct development has evolved in six of these ten orders.

The echinothurioids are unusual in that only direct-developers are known. Evidence suggests that the pluteus is the primitive larval form in the sea urchins. This larva occurs in all orders except echinothurioids. The arms of the pluteus are probably homologous to the noncalcified arms of an aricularia-like primitive echinoderm larva (Ohshima 1911; Kume and Dan 1968; Raff et al. 1988). There is little direct evidence on the evolution of the pluteus. The only fossils are of well-preserved pluteus skeletal rods from the Jurassic of France (Deflande-Rigaud 1946). These appear to be derived from the pluteus of the Jurassic cassiduloid *Echinobrissus scutatus*. Since cassiduloids first appeared in the lower Jurassic, the pluteus evidently was present early in the history of the order. Living cassiduloids include both typical and brooded direct-developing species (Gladfelter 1978; Emlet et al. 1987).

The pluteus has a highly specialized structure, in which ciliated bands on the rigid arms are used for swimming and feeding (Strathmann 1971, 1974, 1975). Morphogenesis of this larva takes place through a complex set of processes, which produce a functional gut, skeleton, arms, musculature, neural systems, and ciliated epithelium. Clearly, if the selective demand for a feeding larva is removed by the evolution of a large yolky egg that provides all the food necessary for development to metamorphosis, feeding structures can be omitted and the genetic programs that produce them lost.

Although plutei are very similar to each other, direct-developing larvae differ considerably (Raff 1987). This is what would be expected if the pluteus is primitive and functionally constrained, but direct-developing larvae are not. All such larvae would have evolved from the pluteus, but by different routes. Since loss of purely pluteus features would be expected to be largely nonselective, extents and modes of loss would be predicted to be highly variable. Direct-developing larvae would also be expected to preserve some of the phylogenetic peculiarities of the plutei from which they arose. Thus, whereas euechinoid plutei and direct-developers have a vestibule, neither cidaroid plutei nor direct-developers have this structure. Novel structures would be selected for, but would differ with details of the new developmental regime.

Given the diversity of direct-developing larvae, and the specialized complexity of the pluteus, if feeding larvae evolved from the various nonfeeding forms, it would be extremely unlikely that all would produce plutei. It is clear from other classes that other styles of feeding larvae are possible for echinoderms (Strathmann 1971, 1974, 1975, 1978). Dollo's law probably applies. Once direct-developing, yolk-rich larvae have evolved and lost the developmental programs for differentiation of feeding structures, it may not be possible for such programs to re-evolve.

One last point should be noted. Examples are known in which two species of the same genus exhibit typical and direct modes of development. This is particularly dramatic in the two species of *Heliocidaris* that we are studying. Molecular data from 18S rRNA sequences (Raff et al. 1988) and from mitochondrial DNA restriction endonuclease digests (current studies in collaboration with Dr. S. Palumbi, University of Hawaii) are consistent with these species being congeneric, although certainly considerably genetically diverged. In many other cases, direct-developers are scattered among families in which the preponderance of species produce typical plutei (Emlet et al. 1987). These occurrences suggest that direct-development can arise readily.

## Conclusions

Direct-developing echinoids, in some cases closely related to species exhibiting typical development, offer accessible experimental systems for examining evolutionary changes in developmental processes at the cellular level. This is important because many of our hypotheses about the role of development in the evolution of metazoan forms are untested and are based on theoretical considerations and interpretation of incidental observations. This is particularly true of ideas about developmental constraints upon changes in early development. Developmental constraints on early development have been proposed because of the general belief that early development is evolutionarily conservative in many groups of organisms. In fact, that conservatism has been the basis for the creation of multiphylum groups such as spiralia or protostomes and deuterostomes. The developmental constraints responsible for this conservatism are believed to result from two basic causes: the inflexibility of developmental mechanisms that prohibit evolutionary changes, and/or deleterious consequences for subsequent developmental steps when early development is altered. However, a major point to be drawn from our studies is that early development may be very loosely constrained and accessible to dramatic evolutionary modifications.

We have demonstrated that homologies can be established between the major cell lineages of typical and direct-developing sea urchin embryos. The establishment of such cellular homologies is a necessary precondition to dissecting the mechanisms that underlie evolutionary changes in developmental mode. Cellular homologies cannot be established on the basis of gene expression patterns alone, although this might be a tempting approach given the availability of probes that recognize cell lineage-specific products. Caution needs to be exercised because both heterochronies and heterotopies have been demonstrated in the expression of such products in cells of various typically developing sea urchins. Thus, we have established a set of guidelines for defining cellular homologies by a variety of approaches including the direct tracing of cell fates by injection of fluorescent dyes into early embryo blastomeres. Once homologies are clearly established, it is possible to dissect the effects of changes in the behavior of individual cell lineages.

Heterochronies are prevalent in the evolution of direct development. The overall modification of the pluteus to produce the direct-developing larva can readily be described in heterochrony terms. This simple model can then be tested against the modified behavior of individual cell lineages. A correspondence between the morphological heterochronies and the program for expression of one cell lineage-restricted gene product has been demonstrated. However, it is clear that a simple heterochrony model for the observed changes in development is too simplistic to serve as more than a very rough approximation of events. This is true for two reasons. First, the very features that allow heterochronous dissociations to occur in early development between cell lineages also generate a mosaic pattern of evolutionary changes in that cell lineages have been differentially responsive to selective pressures. Second, although heterochrony is usually discussed as a mechanism, it is more often a result of mechanisms that are not changes in timing per se. Such nonheterochronous mechanisms have occurred in the evolution of direct development, and have produced heterochronous results in cell cleavage behavior.

Did direct development arise from a "hopeful monster?" Probably not. The complex change in developmental mode more likely arose from several independent alterations. The initial step may well have been the achievement of a large egg, which develops into a nonfeeding pluteus. Once this occurred, whole suites of other modifications of larval development followed because the dependence on feeding structures had been removed. Other selective pressures, including those favoring an acceleration of adult development, could result both in loss of pluteus features and in gain of novel features associated with a new mode of larval life and rapid adult development. The frequency with which direct development has arisen independently in several sea urchin groups indicates that these changes are not difficult to achieve genetically or developmentally.

Finally, we should consider whether the changes in *H. erythrogramma* development constitute evolutionary innovations. The question arises because other authors in this volume have defined innovations as key features which permit the radiation of some group of organisms possessing the innovation. From this perspective, innovations are recognized by their historical consequences in speciation events. Without taking issue with this approach to the definition of innovation, we as developmental biologists dealing with living systems necessarily see innovation in a different light. We are concerned with the detailed developmental mechanisms by which novel structures evolve without immediate regard to selective forces or ultimate evolutionary results. Thus, novel larval features that can be understood at cellular and genetic levels allow a dissection of the means by which developmental programs can evolve. These are the raw materials for the spectacular parade of evolutionary innovations seen in the fossil record and around us in the living world.

## Acknowledgments

These studies were supported by research grant HD21337 from NIH, an NSF US/Australia Cooperative Program travel grant, and a Guggenheim Fellowship to R.A.R. We thank Donald T. Anderson and Valerie M. Morris, University of Sydney, for their collaboration in this research, and Ian Hume, Chairman of the Department of Zoology at the University of Sydney, for generously making the facilities of the department available to us.

## References

Akam, M. 1987. Molecules and morphology. *Nature* 327: 184-85.

Alberch, P. 1982. Developmental constraints in evolutionary processes, pp. 313-32. In *Evolution and Development,* ed. J. T. Bonner. Berlin: Springer-Verlag.

Alberch, P. 1985. Problems with the interpretation of developmental sequences. *Systematic Zoology* 34: 46-58.

Alberch, P. 1987. Evolution of a developmental process: Irreversibility and redundancy in amphibian metamorphosis, pp. 23-46. In *Development as an Evolutionary Process,* ed. R. A. Raff and E. C. Raff. New York: Alan R. Liss.

Anstrom, J. A., J. E. Chin, D. S. Leaf, A. L. Parks, and R. A. Raff. 1987. Localization and expression of msp130, a primary mesenchyme lineage-specific cell surface protein of the sea urchin embryo. *Development* 101: 255-65.

Beeman, R. W. 1987. A homoeotic gene cluster in the red flour beetle. *Nature* 327: 247-49.

Bock, W. J. 1974. Philosophical foundations of classical evolutionary classification. *Systematic Zoology* 22: 375-92.

Buss, L. W. 1987. *The Evolution of Individuality.* Princeton: Princeton University Press.

Cameron, R. A., B. R. Hough-Evans, R. J. Britten, and E. H. Davidson. 1987. Lineage and fate of each blastomere of the eight-cell sea urchin embryo. *Genes and Development* 1: 75-84.

Czihak, G. 1960. Untersuchungen uber die Coelmanlagen und die Metamorphose des Pluteus von *Psammechinus miliaris* (Gmelin). *Zoologische Jahrbucher.* Abt II: *Anatomie and Ontogenie der Tiere* 78: 235-56.

Czihak, G. 1965. Entwicklungsphysiologische Untersuchungen an Echiniden. Experimentelle Analyse der Coelomentwicklung. *Wilhelm Roux's Archive Entwicklungsmechanics Organismen* 155: 709-29.

Davidson, E. H. 1986. *Gene Activity in Early Development.* 3d ed. New York: Academic Press.

de Beer, G. 1958. *Embryos and Ancestors.* 3d ed. Oxford: Oxford University Press.

de Beer, G. 1971. Homology, an unsolved problem, pp. 3-11. In *Oxford Biology Readers,* ed. J. J. Head and O. E. Lowenstein. London: Oxford University Press.

Deflande-Rigaud, M. 1946. Vestiges microscopiques des larves d'Échinoderms de l'Oxfordien de Villers-sur-Mer. *Comptes Rendus des Séances de l'Académie des Sciences* 222: 908-10.

Elinson, R. P. 1987. Change in developmental patterns: Embryos of amphibians with large eggs, pp. 1-21. In *Development as an Evolutionary Process,* ed. R. A. Raff and E. C. Raff. New York: Alan R. Liss.

Emlet, R. B. 1986. Facultative planktotrophy in the tropical echinoid *Clypeaster rosaceus* (Linnaeus) and a comparison with obligate planktotrophy in *Clypeaster subdepressus* (Gray) (Clypeasteroidea: Echinoidae). *Journal of Experimental Marine Biology and Ecology* 95: 183-202.

Emlet, R. B., L. R. McEdward, and R. R. Strathmann. 1987. Echinoderm larval ecology viewed from the egg, pp. 55-136. In *Echinoderm Studies 2,* ed. J. Lawrence and M. Jangoux. Rotterdam: Balkema.

Garstang, W. J. 1928. The origin and evolution of larval forms. *Report of the British Association for the Advancement of Science, Sec. D.*

Gladfelter, W. B. 1978. General ecology of the cassiduloid urchin *Cassidulus caribbearum. Marine Biology* 47: 149-60.

Gould, S. J. 1977. *Ontogeny and Phylogeny.* Cambridge: Harvard University Press.

Gould, S. J. 1982. Change in developmental timing as a mechanism of macroevolution, pp. 333-46. In *Evolution and Development,* ed. J. T. Bonner. Berlin: Springer-Verlag.

Horstadius, S. 1973. *Experimental Embryology of Echinoderms.* London: Oxford University Press.

Hursh, D. A., M. E. Andrews, and R. A. Raff. 1987. A sea urchin gene encodes a polypeptide homologous to epidermal growth factor. *Science* 237: 1487-90.

Kume, M., and K. Dan. 1968. *Invertebrate Embryology.* Translation of the original published by Bai Fu Kan Press, Tokyo, 1957. Belgrade: NOLIT.

Leaf, D. S., J. A. Anstrom, J. E. Chin, M. A. Harkey, R. M. Showman, and R. A. Raff. 1987. Antibodies to a fusion protein identify a cDNA clone encoding msp130: A primary mesenchyme specific cell surface protein of the sea urchin embryo. *Developmental Biology* 121: 29-40.

Maynard Smith, J., R. Burian, S. Kauffman, P. Alberch, J. Campbell, B. Goodwin, R. Lande, D. Raup, and L. Wolpert. 1985. Developmental constraints and evolution. *Quarterly Review of Biology* 60: 265-87.

Mortensen, T. 1928. *A Monograph of the Echinoidea. I. Cidaroidea.* Copenhagen: C. A. Reitzel.

Mortensen, T. 1935. *A Monograph of the Echinoidea. II. Bothriocidaroida, Melonechinoida, Lepidocentroida, and Stirodonta.* Copenhagen: C. A. Reitzel.

Mortensen, T. 1940. *A Monograph of the Echinoidea. III.1 Aulodonta.* Copenhagen: C. A. Reitzel.

Mortensen, T. 1943a. *A Monograph of the Echinoidea. III.2 Camarodonta. I.* Copenhagen: C. A. Reitzel.

Mortensen, T. 1943b. *A Monograph of the Echinoidea. III.3 Camarodonta. II.* Copenhagen: C. A. Reitzel.

Mortensen, T. 1948a. *A Monograph of the Echinoidea. IV.1 Holectypoida, Cassiduloida.* Copenhagen: C. A. Reitzel.

Mortensen, T. 1948b. *A Monograph of the Echinoidea. IV.2 Clypeastroida.* Copenhagen: C. A. Reitzel.

Mortensen, T. 1950. *A Monograph of the Echinoidea. V.1 Spatangoida. I.* Copenhagen: C. A. Reitzel.

Mortensen, T. 1951. *A Monograph of the Echinoidea. V.2 Spatangoida. II.* Copenhagen: C. A. Reitzel.

Needham, J. 1933. On the dissociability of the fundamental process in ontogenesis. *Biological Reviews* 8: 180-223.

Newport, J., and M. Kirschner. 1982a. A major developmental transition in early *Xenopus* embryos: I. Characterization and timing of cellular changes at the midblastula stage. *Cell* 30: 675-86.

Newport, J., and M. Kirschner. 1982b. A major developmental transition in early *Xenopus* embryos: II. Control of the onset of transcription. *Cell* 30: 687-96.

Ohshima, H. 1911. Larva of echinoderms (in Japanese). *Zoological Magazine* 23: 377-94.

Okazaki, K. 1975. Normal development to metamorphosis, pp. 177-232. In *The Sea Urchin Embryo,* ed. G. Czihak. Berlin: Springer-Verlag.

Parks, A. L., B. A. Parr, J. E. Chin, D. S. Leaf, and R. A. Raff. 1988. Molecular analysis of heterochronic changes in the evolution of direct developing sea urchins. *Journal of Evolutionary Biology* 1: 27-44.

Raff, R. A. 1987. Constraint, flexibility, and phylogenetic history in the evolution of direct development in sea urchins. *Developmental Biology* 119: 6-19.

Raff, R. A. 1988. Direct development in sea urchins: A system for the study of developmental processes in evolution, pp. 63-69. In *Proceedings of the 6th International Echinoderm Conference, Victoria, British Columbia,* ed. R. Burke, P. Mladenov, P. Lambert, and R. Parsley. Rotterdam: Balkema.

Raff, R. A., J. A. Anstrom, C. J. Huffman, D. S. Leaf, J.-H. Loo, R. M. Showman, and D. E. Wells. 1984. Origin of a gene regulatory mechanism in the evolution of echinoderms. *Nature* 310: 312-14.

Raff, R. A., K. G. Field, M. T. Ghiselin, D. J. Land, G. J. Olsen, A. L. Parks, B. A. Parr,

N. R. Pace, and E. C. Raff. 1988. Molecular analysis of distant phylogenetic relationships in echinoderms, pp. 29-41. In *Echinoderm Phylogeny and Evolutionary Biology*, ed. C. R. C. Paul and A. B. Smith. Oxford: Oxford University Press.

Raff, R. A., and T. C. Kaufman. 1983. *Embryos, Genes, and Evolution.* New York: Macmillan.

Raff, R. A., and E. C. Raff, eds. 1987. *Development as an Evolutionary Process.* New York: Alan R. Liss.

Raff, R. A., and G. A. Wray. N. d. Heterochrony: Developmental mechanisms and evolutionary results. Submitted.

Remane, A. [1952] 1971. *Die Grundlagen des Natürlichen Systems der Vergleichenden Anatomie und der Phylogenetik.* 2d ed. Authorized reprint of 1st edition. Königstein-Taunus: Koeltz.

Riedel, R. 1978. *Order in Living Organisms.* Chichester: John Wiley.

Roth, V. L. 1984. On homology. *Biological Journal of the Linnaean Society* 22: 13-29.

Roth, V. L. 1988. The biological basis of homology. In *Ontogeny and Systematics*, ed. C. S. Humphries. New York: Columbia University Press.

Sedgwick, A. 1894. On the law commonly known as von Baer's; and on the significance of ancestral rudiments in embryonic development. *Quarterly Journal of Microscopic Science* 36: 35-52.

Seeger, M. A., and T. C. Kaufman. 1987. Homoeotic genes of the Antennapedia Complex (ANT-C) and their molecular variability in the phylogeny of the Drosophilididae, pp. 179-202. In *Development as an Evolutionary Process*, ed. R. A. Raff and E. C. Raff. New York: Alan R. Liss.

Stent, G. S. 1985. The role of cell lineage in development. *Philosophical Transactions of the Royal Society of London* B312: 3-19.

Sternberg, P. W., and H. R. Horvitz. 1981. Gonadal cell lineages of the nematode *Panagrellus redivivus* and implications for evolution by modification of cell lineage. *Developmental Biology* 88: 147-66.

Strathmann, R. R. 1971. The feeding behavior of planktotrophic echinoderm larvae: Mechanisms, regulation, and rates of suspension feeding. *Journal of Experimental Marine Biology and Ecology* 6: 109-60.

Strathmann, R. R. 1974. Introduction to function and adaptation in echinoderm larvae. *Thalassia Jugoslavia* 10: 321-39.

Strathmann, R. R. 1975. Larval feeding in echinoderms. *American Zoologist* 15: 717-30.

Strathmann, R. R. 1978. The evolution and loss of feeding larval stages of marine invertebrates. *Evolution* 32: 894-906.

Wells, D. E., J. A. Anstrom, R. A. Raff, S. R. Murray, and R. M. Showman. 1986. Maternal stores of α subtype histone mRNAs are not required for normal early development of sea urchin embryos. *Roux's Archives of Developmental Biology* 195: 252-58.

Williams, D. H. C., and D. T. Anderson. 1975. The reproductive system, embryonic development, larval development and metamorphosis of the sea urchin *Heliocidaris erythrogramma* (Val.) (Echinoidea: Echinometridae). *Australian Journal of Zoology* 23: 371-
403.

Wilson, E. B. 1894. The embryological criterion of homology. *Biological Lectures Delivered
at the Marine Biological Laboratory of Woods' Holl [Marine Biological Laboratory,
Woods Hole, Massachusetts].* 1894: 101-24.

Wray, G. A., and D. R. McClay. In press (a). The origin of spicule-forming cells in a "primitive" sea urchin (*Eucidaris tribuloides*), which appears to lack primary mesenchyme
cells. *Development.*

Wray, G. A., and D. R. McClay. In press (b). Molecular heterochronies and heterotopies in
early echinoid development. *Evolution.*

Wray, G. A., and R. A. Raff. In press. Evolutionary modifications of cell lineage and fate in
the direct-developing sea urchin *Heliocidaris erythrogramma*. *Developmental Biology.*

# Developmental Mechanisms
# at the Origin of Morphological Novelty:
# A Side-Effect Hypothesis

*Gerd B. Müller*

> There are certain periods when things are profoundly transformed of their
> own accord, assuming new forms that develop of themselves, rather like
> crystals.
> —André Malraux

Novelty, the appearance of a new structural feature, is a rare phenomenon
in morphological evolution compared to the immense number of changes
that are realized through modifications of size, proportion, and shape. Nevertheless, it represents a fundamental problem of evolutionary theory – chal-
lenging the neo-Darwinian dogma – how new structures can appear at all in
organisms, because such characters evidently cannot be selected for before
they come into existence. Development and its mechanisms are unquestion-
ably central to the problem of novelty, since phylogenetic changes of morphology necessarily require modifications of ontogeny. On the other hand,
recent studies indicate a lack of close correlation between genetic and morphological evolution (Wilson et al. 1977; Wake 1981; Larson et al. 1984;
John and Miklos 1988), suggesting that alterations of the genome are to
some extent peripheral to the problem of morphological change. This is
supplemented by an increasing number of studies (Alberch and Gale 1983,
1985; Brylski and Hall 1988a, 1988b; Sinervo and McEdward 1988; Müller
1986, 1989; Müller and Streicher 1989) which provide evidence that the historically acquired and lineage-specific properties of developmental systems

have a dominating influence on evolutionary modifications of morphology. For this reason it is desirable to analyze novelty from a developmental point of view in contrast to earlier discussions that centered on selectionist genomic scenarios (Mayr 1976).

In organismic evolution we can distinguish two categories of morphological novelty. One is the generation of entire new body plans at the origin of the major taxonomic groups, the other comprises novelties that result from the restructuring and transformation of existing body plans during the diversification of a class of organisms. Both types of novelty must be realized through the alteration of developmental programs. Since the basic topography of body design is laid out very early in ontogeny, we may expect major new body plans to result from modifications at early stages of development, such as cleavage and gastrulation, in the possibly less specialized ontogenies of primitive organisms. This is supported by the fact that, with the possible exception of bryozoans, no major new body plans appeared after the Cambrian, making it also problematic to approach this category of novelty empirically. Despite this difficulty, useful hypotheses have been proposed (Garstang 1922; Severtzoff 1931; Dalcq 1949; Raff and Kaufman 1983; John and Miklos 1988), and recent experimental studies show that change in egg size alone has profound effects on larval form (Sinervo and McEdward 1988). Other aspects of modifications in early development are discussed elsewhere (Raff et al., this volume).

The present essay will concentrate on the second category – novelties that result from the modification of particular organ systems within a given body plan. Novelties of this type consist of the appearance of entirely new structures that were not present in the ancestral group, such as the maxillary joint in bolyeriid snakes, the cheek pouches of geomyoid rodents, the fibular crest on the theropod tibia, the giant panda's additional digits, the narwhal's tusk, or the turtle carapace. In contrast to entire new body plans, most of these organ-level innovations are realized at later stages of development, during the morphogenesis of tissues and organs, making the category also accessible to experimental studies of the mechanisms involved.

An immanent difficulty of the subject is the lack of a clear definition of morphological novelty. A definition could include taxonomic criteria since the appearance of novelty is often associated with speciation, but there is not a necessary correlation between the two (Mayr 1976). Further, new functional properties have been suggested as indicators of novelty, especially in the case of key evolutionary innovations (Liem, this volume), but it was also noted that a change of function characteristically precedes the alteration of a structure (Mayr 1976). In general, novelty seems to indicate a departure from the continuous and gradual evolutionary alterations of morphology. This means that the term does not apply to mere quantitative changes but always indicates qualitative modifications of morphology. Thus novelty is here defined as a qualitatively new structure with a discontinuous origin, marking a relatively abrupt deviation from the ancestral condition. Examples such as those mentioned above meet these requirements and will provide the basis of my analysis.

Before discussing mechanisms of development that can underlie the generation of new organismic structures, I will briefly consider the conceptual frame for relating processes of development to evolution and novelty. I will then propose three properties of developing systems – thresholds, intermediate structures, and switching of mechanisms – potential generative sources of novelty. A fourth property, developmental plasticity, will be discussed as an essential requirement to facilitate the integration of new structures into the organism. Finally, I will consider the consequences of a developmental approach to novelty for evolutionary theory.

## Concepts and Open Questions

"Embryology is to me by far the strongest single class of facts in favor of a change of forms," Darwin noted in a letter to Asa Gray, and later com-

plained on several occasions about the neglect of his embryological arguments (Mayr 1982). This paradox became symbolic for the position of embryology in evolutionary theory, especially after the rise and fall of Haeckel's
doctrine. Waddington, for example, expressed on many occasions (e.g., 1941)
that "evolutionary theory requires a picture of the possible interactions between developmental processes." Stebbins (1968) regards "the nature of epigenetic sequences of vital importance to students of evolution," and numerous other quotations to that effect are found in the literature. The prominent position assigned to embryology in such statements, as well as in more
elaborate treatments (e.g., de Beer 1938; Goldschmidt 1940; Schmalhausen
1949; Waddington 1957, 1962; Bertalanffy 1952), contrasts sharply with the
actual role development assumed in the evolutionary synthesis. Historically,
embryology has neither contributed to the establishment of the synthesis
(Hamburger 1980; Mayr 1980), nor has it thereafter become an important
conceptual issue (Dobzhansky et al. 1977; Maynard Smith 1982; Sober 1986).
Recently, however, it has again been suggested that alterations in the epigenetic system may play an important role in macroevolution (Alberch 1980;
Hall 1983; Ho and Saunders 1979; Sinervo and McEdward 1988). While
most of these assertions come from developmental biologists, a number of
theoretical population-genetic analyses, which demonstrate the adaptive
advantage of epigenetically coupled characters (Wagner 1984; Bürger 1986),
lend further support to this notion.

Although embryology had not much influence on the synthetic theory,
there existed a continuous interest in the relationship between development
and evolution. As a consequence three major concepts emerged: recapitulation, heterochrony, and constraint. Neither one, however, provides a mechanistic explanation for the origin of novel structures. Recapitulation is mainly a phenomenological concept about the parallels between sequences of
development and those of phylogeny (see Gould 1977 for a review). The
concept of developmental constraint (Alberch 1982; Maynard Smith et al.
1985) relates to the limitations of phenotypic variability that result from the

properties of developmental systems. Only heterochrony, the phylogenetic modification of the rates and timing of developmental processes, represents a concept dealing with the mechanisms that can alter ontogenies. A variety of developmental parameters, such as mitotic rates, growth rates, inductive interactions, etc., can be affected by heterochrony (Hall 1984). However, while modifications of developmental timing possibly represent a dominant mode of morphological evolution (e.g., de Beer 1958; Gould 1977, 1982; Alberch et al. 1979; McNamara 1982; Raff et al., this volume), these modifications can only concern processes that already exist in ontogeny. Since it is confined to the modification of the existing, heterochrony alone cannot explain the origin of new developmental pathways and the generation of specific novelty. Thus we are faced with a variety of open questions. If heterochrony is a major mode of morphological evolution, what is its relationship to novelty? Is it necessary to evoke additional mechanisms that are active in the generation of novelty? What determines the specificity of morphological innovations? Is it possible to identify particular properties of development that can promote the generation of novelty and can they account for the apparent rapidity in the origin of novel characters? The following sections attempt to answer some of these questions by analyzing four properties of developmental systems that may represent fundamental factors in the origin of novelty.

## Thresholds in Developing Systems

To propose developmental threshold effects as one key factor in the generation of novelty is based on the fact that most evolutionary changes of morphology are associated with modifications of size and proportions of body parts or whole organisms. We must assume that selection for size, which is essentially selection for changes in developmental rate, can eventually push the developing cell or tissue system concerned to the boundaries of its existing steady state appropriate for a particular morphology. Boundaries of steady state systems usually are not continuous but have threshold qualities. The system will assume a new steady state upon the crossing of the threshold and the resulting phenotypic transformation will then depend on the reaction norms of the system at this point, as well as on the secondary reactions of associated systems.

It was first recognized by Rensch (1948) that evolutionary changes in size often correlate with the histological alteration of structure. Using examples from the flight apparatuses, the nervous systems, and the intestinal tracts of closely related insects, he demonstrated that a continuous increase of body size, but also its extreme decrease, obligatorily leads to new structures at the histological level (Rensch 1948). Rensch points out that natural selection primarily acts on body size and only secondarily on single organs and their histological composition, but he does not make a connection to the causal mechanisms of their development. Some examples from more recent experimental studies in vertebrates illustrate how changes of size can affect mechanisms of pattern formation and morphogenesis through the crossing of developmental thresholds.

First I will consider possible effects of body size changes on early organ blastemata and their consequences on pattern. In skeletal development it is known that a critical number and density of prechondrogenic cells is necessary to start the expression of cartilage matrix and the formation of a chondrogenic condensation (Ede et al. 1977; Newman 1977; Solursh 1984). Based on observations of cultured aggregates, Ede and Flint (1972) suggest that a simple threshold response to a carbon-dioxide/oxygen gradient may initiate the condensation act. The number of chondrogenic centers that form within a limb also relates to the geometry of the limb bud (Oster et al. 1988), especially to its width, as demonstrated experimentally and by the *talpid*[3] mutant of the chick (Hinchliffe and Johnson 1980) (fig. 1). These properties of early skeletogenesis suggest that an increase or decrease in mesenchymal cell number or in overall size of limb buds can reach upper or

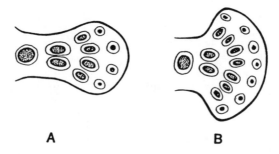

Figure 1.  Schematic interpretation of precartilage fields in normal (A) and *talpid*[3] (B) limb buds of the chick.  After Hinchliffe and Johnson 1980.

lower thresholds at which the number of condensations and consequently the number of skeletal elements in the adult limb will change.

A variety of investigations that experimentally alter mesenchymal cell number in vertebrate limb buds support a threshold notion.  Reduction of cell number at prechondrogenic stages through X-ray irradiation (Wolff and Kieny 1962; Wolpert et al. 1979) or through antimitotic drugs (Bretscher 1949; Bretscher and Tschumi 1951; Raynaud 1985) does not produce limbs that are harmonically reduced in size but often results in the absence of specific skeletal elements and in digital reduction.  Bretscher and Tschumi (1951) demonstrated the threshold phenomenon underlying such effects in a large series of drug-treated *Xenopus* limbs, which fall into discrete classes of digital reduction but almost never show intermediate forms (fig. 2).  The evolutionary relevance of such experimental patterns has been convincingly shown by Alberch and Gale (1983, 1985).  They not only induced phalangeal reduction and digital loss in amphibian limbs by application of a mitotic inhibitor, but they were also able to show that the rearrangements in skeletal morphology parallel the phylogenetic trends of digital reduction in frogs and salamanders respectively.

In theory, complementary effects are expected to result from an increase of cell number in limb buds.  While this has not yet been demonstrated experimentally, the more frequent occurrence of additional digits in

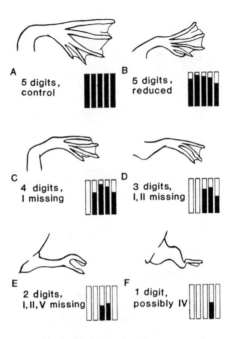

Figure 2. Digital reduction in *Xenopus* hindlimbs following colchizin treatment of the limb buds. Rather than producing harmonic reduction of the whole foot, entire digits disappear in the sequence I, II, V, III (C-F). Length reduction of the remaining digits is indicated by the bar graphs. After Bretscher 1947.

large dog species (Alberch 1985) as opposed to small species does seem to fit the picture. It remains to be established, however, whether limb bud size and cell number are actually smaller in embryos of smaller species. So far we may conclude that the evolutionary reduction of blastema size can affect final limb patterns via the threshold of critical cell number necessary to initiate skeletogenesis. Developmentally, cell number in the limb primordium may be altered not only through changes in the initial size of the limb bud, but also by the immigration of additional cells, or through secondary cell death (Hinchliffe and Johnson 1980).

I now turn to another aspect of embryogenesis in which subtle continuous variation of proportions can induce morphological change through the crossing of a threshold. This is the relative position of an organ primordium and its interaction with adjacent structures. Differential rates of development, for example, can lead to changes of position and interaction of formerly separate tissues, with the potential to initiate an immediate morphological discontinuity. An attractive example of this kind has recently been presented by Brylski and Hall (1988a, 1988b), who analyzed the ontogeny of external and internal cheek pouches in rodents. Some groups of rodents possess pouches which open internal to the mouth and are lined with buccal epithelium. Others, namely the Geomyidae and Heteromyidae, are characterized by fur-lined pouches which open external to the mouth and have no connection to the mouth cavity itself. As internal pouches are thought to be primitive, the external pouch type presents a classic case of novelty for which a discontinuous origin is claimed (Long 1976). Brylski and Hall (1988a) observed that both types of pouches start their development from an invagination of the buccal epithelium, close to the lateral corner of the mouth, but only in the formation of external pouches does the invagination eventually include the lip epithelium of the mouth corner itself, initiating the externalization of the pouch. The authors point out that it took only a minimal anterior shift of the invagination area to produce a quite dramatic phenotypic change, in which furring of the pouch could be the immediate result of the interaction of buccal epithelium with dermal mesenchyme. Externalization of the invaginated epithelium coincides with the elongation of the snout, suggesting that slight allometric changes in facial development ultimately caused the shift of the invagination site and the inclusion of the lip.

A most interesting aspect of this scenario is that, based on the pouch ontogeny, Brylski and Hall (1988b) see no possibility for an intermediate phenotype, i.e., a pouch with both internal and external openings, suggesting a rapid evolutionary transition from one to the other. The sudden appearance in the fossil record of a full turtle carapace (Carroll 1988), recently proposed to have evolved through a similarly small modification of epithelial-mesenchymal interactions (Burke 1989), also seems to leave little possibility for intermediate phenotypes.

The biomechanics of embryonic development are the basis on which a third category of threshold effects can occur. Several developmental processes, such as chondrogenesis or osteogenesis, are known to respond strongly to mechanical stimulation (Hall 1978, 1979). It is evident that evolutionary modifications of the size or proportions of a structure not only have effects on the mechanical properties of the adult but also on those of the developing embryo. Changes in tension, pressure, shear, movement, etc., can easily generate a modified tissue response once an adequate threshold is reached. Thus it is possible to initiate or suppress secondary cartilage formation experimentally by modifying biomechanical conditions in dermal bone (Hall 1970, 1986). Tendons and ligaments have a similar capacity to respond to mechanical influences by forming cartilage in the stress area (Hall 1978; Merrilees and Flint 1980). As part of an evolutionary scenario it was recently proposed that the formation of a sesamoidlike cartilage between the tibia and the fibula of developing bird limbs resulted from the increasing mechanical instability of the embryonic fibula during its progressive evolutionary reduction, intensifying stress on the ligamentous connection between the two bones (Müller and Streicher 1989). The role of biomechanics in the ontogeny of this character can be tested by paralysis experiments that cause a failure of the cartilage element to form (Müller and Hall, unpublished). In this context it is important to understand that embryonic movement starts very early in the ontogeny of birds and maintains a high intensity for an extensive period of time (Hamburger et al. 1965).

The links between thresholds at the developmental level and evolutionary changes of body size are indicated by many of the experimental studies mentioned above, but can also be derived from statistical analyses of the correlation of body size with morphological change in comparative studies. Various genera of Scincid and Teiid lizard families, for example, show progressive body elongation in combination with limb reduction. Structural reduction of the limbs was shown to begin at a group-specific threshold of relative limb size (Lande 1977). Lande also noted that a second threshold exists for the total loss of digits when relative limb reduction attains less than half snout-eye length. Characteristically, streamlining and elongation of the body preceded limb reduction in lizards (Gans 1975). Other studies document the high correlation of miniaturization with the occurrence of novel morphologies (Hanken 1985).

The list of threshold phenomena in embryogenesis can be extended further, including thresholds at various levels of development, such as in molecular interactions (hormones, morphogens, adhesion molecules, etc.), in inductive tissue interactions, in temperature-dependent processes, and in various other domains of developmental physiology. This essay, however, does not intend to provide an exhaustive list of such effects; rather, it attempts to emphasize that thresholds are an inherent property of developing systems, able to trigger discontinuities in morphogenesis which can automatically result in the generation of a new structure. Novelty can thus arise as a side effect of evolutionary changes of size and proportion, with the specific result depending on the reaction of the affected systems. In this scenario the emerging structure becomes only secondarily a target of selection which will determine its maintenance and persistence throughout the population; the disruption of a morphogenetic sequence lies at its origin. Under this evolutionary heading, the study of thresholds in development should receive more empirical attention.


## Intermediate Structures in Ontogeny


Threshold effects in development are not necessarily immediately expressed in the adult phenotype but can initially produce a transitory structure introduced into the course of ontogeny. A great number of other structures also exist only during certain stages of embryonic development and are no longer present in the adult. In vertebrates the best known examples for such transitory structures come from the aortic arch system, the progressive nephric series in kidney development, and from skeletal development, but the same is true for other organ systems and for nonvertebrate ontogenies.

Ontogenetically temporary or intermediate structures were called "interphenes" by Riedl (1978), who also discusses their origin and role in development. Essentially two kinds of interphenes can be distinguished. One is those structures that are truly recapitulatory remnants of the ontogenetic history of the lineage – palingenetic structures of Haeckel (1866). The second kind comprises structures that evolved for the maintenance of embryonic (larval) life, or as a consequence of particular embryonic conditions that no longer exist in the adult – called caenogenetic structures by Haeckel (1866), who also included structures that arise through heterochrony or heterotopy into this category. In addition it might be possible to conceive of a third kind of intermediate structures that are neither recapitulatory nor adaptations to embryonic conditions, but which are neutral by-products of developmental processes tolerated by the system: a developmental analogy to neutrality at the genome level. I will propose in this section that intermediate structures represent an important ontogenetic source for novelty in morphological evolution. I shall discuss two examples in support of this hypothesis, one for each of the two categories of interphenes.

Many readers will be familiar with the cases of *Bolyeria* and *Casarea*, two genera of bolyeriid snakes from the Round Islands near Mauritius, which are frequently quoted in the context of morphological novelty in evolution. These snakes are unique in that they are the only terrestrial vertebrates to possess a movable joint between the anterior and posterior parts of the maxilla, a feature that seems to have arisen discontinuously in the phylogeny of the group (Frazzetta 1970). Although the head development of these two snakes has not been studied, it is known from the colubridae, a sister group of the bolyeriids, that ossification does not proceed uniformly in the maxilla, but starts from several foci which are eventually concentrated into two distinct centers of ossification (Haluska and Alberch 1983). The central portion between the anterior and the posterior well-ossified parts of the maxilla does not calcify completely until the 59-day stage of their ontogeny.

This observation corroborates Anthony and Guibe's (1952) suggestion that the bipartid maxilla in bolyeriid snakes may result from a failure of the two centers of ossification to fuse, a notion that is further supported by comparative and embryological analyses of other snake taxa (Irish and Alberch, unpublished). Under the fairly safe assumption that the mechanism of ossification in bolyeriid skulls is essentially similar to the one in the sister taxa, it is possible to conclude that the intramaxillary joint in bolyeriids arose through truncation of the ossification process and retention of the two separate bones by the adult, i.e., through paedomorphic heterochrony. Thus we have a case in which the suppression of the last step of a developmental sequence results in the elaboration and phenotypic expression of a formerly transitory ontogenetic stage – novelty, based on a palingenetic condition.

An example of a caenogenetic character serving as the basis for the evolution of a new structure comes from our own studies of dinosaur and bird hindlimbs (Müller and Streicher 1989). The novelty, in this case, consists of a large osseous crest on the tibia of theropod dinosaurs, providing an unusual link between the tibia and the fibula. No other group of dinosaurs or reptiles is known to possess this feature, but it is synapomorphic for birds, which are now commonly thought to have descended from the theropods (Ostrom 1976; Feduccia 1980). The "fibular crest" on the bird tibia (tibiotarsus) is an important part of a syndesmotic joint between the two bones, essential for the biomechanics of the bird hindlimb. It is, of course, not possible to observe directly the ontogeny of the dinosaur limb to determine the developmental mechanisms involved in the production of the crest, but we can observe its development in the bird embryo. There is enough evidence from reptilian embryology and comparative anatomy to infer that all these mechanisms already existed in dinosaurs and there is no reason to assume that the structure is formed in a fundamentally different manner in birds (Müller and Streicher 1989).

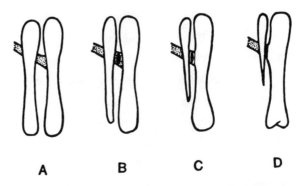

Figure 3. Ontogeny of the syndesmosis tibiofibularis in chick limbs. A cartilaginous sesamoid appears in the tendinous connection between tibia and fibula (B). The sesamoid induces crest formation on the tibia (C) and is eventually fully replaced by the osseous crest (D). After Müller and Streicher 1989.

In chick ontogeny the fibular crest does not start to form as a simple osseous outgrowth from the tibia. Initially an isolated cartilage element appears within the connective tissue precursor of the syndesmosis (fig. 3). The element functions as a sesamoid during embryonic movement, and increases dramatically in size before any crest formation begins on the tibia. The capacity of tendons to develop sesamoids as a response to local mechanical stress was discussed in the threshold section. Through its reactive origin and the limitation of its function to the developmental period, the cartilage sesamoid qualifies as a caenogenetic structure. The normal ontogeny of the later-forming osseous fibular crest is dependent on the presence of the cartilage element, which stimulates bone growth on the tibia and finally becomes incorporated into the crest, leading to the complete disappearance of the cartilage element itself. Assuming that a homologous mechanism of crest ontogeny existed in the dinosaur limb, the novelty "fibular crest" was formed on the basis of a transitory embryonic structure, the cartilage sesamoid, which itself likely originated as a consequence of increased biomechanical stress in the area (see Müller and Streicher 1989 for a detailed discussion).

While at first this seems like a singular example, it can be shown that a series of other well-known cases of skeletal novelty apparently was gener-

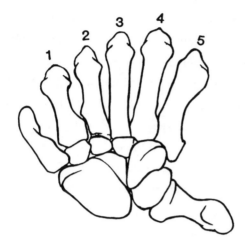

Figure 4. Carpus and metacarpus of the giant panda. The metacarpals homologous to the pentadactyl vertebrate hand are numbered 1 to 5. Note the additional "digits" on either side of the hand. After Davis 1964.

ated by similar mechanisms. The famous "thumb" of the giant panda (Davis 1964; Gould 1980), a small additional "digit" on the radial side of the hand, is based on the enlargement of a sesamoid, and so is the mostly forgotten additional "seventh digit" on the ulnar side of its hand (fig. 4). The two "digits" arose from intratendinous cartilages and not from the modification of digital pattern-forming processes, which also explains the absence of distal phalangeal elements in both. Although we do not know the exact embryology of these structures, they represent remarkable examples of integrated novelty, especially as a new set of muscles that render the "digits" movable and functional is also present (Davis 1964). Many other innovations in the skeletal system of vertebrates may also be based on mechanisms of stress-induced chondrogenesis or osteogenesis, especially structures of "dermal bone," such as the clavicle or the furcula in birds (Hall 1975).

The characteristic of ontogenetic systems to pass through sequences of temporary structures provides a plethora of opportunities for the formation of novelty. Intermediate structures can be retained, incorporated, modified, amplified, etc., mainly triggered through heterochronic changes in development or through threshold events. The presence of such "pre-existing" structures in ontogeny may account for the rapidity and discontinuity of some phylogenetic changes in phenotypes, as often observed in the fossil record. Preadaptation may, therefore, be considered a phenomenon that frequently resides in the developmental properties of an organismic lineage, rather than in the functional conditions of their adult phenotypes. This applies not only to embryonic structures but also to the mechanistic potentials of ontogeny. In the context of novelty it is particularly important to consider the succession of mechanisms during morphogenesis and the extension of pre-existing mechanisms to new areas.

### Switching of Developmental Mechanisms

The phases of embryonic organ formation are not only distinguished by increasing levels of complexity but also by the changes in the developmental mechanisms governing each stage. A characteristic example is the development of the vertebrate limb skeleton with consecutive stages of pattern formation: prechondrogenic condensations, chondrogenic condensations, cartilage rudiments, and ossification. The switching from one mechanism to the next each time holds opportunities for qualitative morphological change. Nowhere is this more apparent than in the repatterning during amphibian metamorphosis, but, as metamorphosis is a very particular process, I will exemplify the evolutionary ramifications of the switching aspect using cases from vertebrate limb evolution.

The early skeletal patterning process in the vertebrate limb bud is clearly a very conservative process. Shubin and Alberch (1986) and Oster et al. (1988) have proposed a mechanistic model of embryonic branching and segmentation in initial chondrogenesis. The model provides a morphogenetic explanation for the conservativism of these primary patterns across the tetrapod taxa. It also shows that modifications of the skeletal patterns already occur during these early phases, mainly through delays and truncations in the branching and segmentation process. Many evolutionary modifications of the limb skeleton can, therefore, reside in very early heterochronic alterations of the primary pattern-forming processes. However, a large number of other phylogenetically important transformations and innovations are achieved through modifications of the primary patterns during later stages of ontogeny. These are often reductions of the initial number of skeletal elements through either deletions or fusions.

Fusions typically occur in association with the takeover of a new mechanism during skeletogenesis. One such phase is the transition from prechondrogenic cell aggregations to cartilage matrix expression resulting in well-defined cartilage rudiments. In many cases alterations in the number of carpal and tarsal elements are based on such early fusions. In the alligator embryo the condensation distal to the radius (as part of the preaxial ray) and the condensation of the intermedium (as part of the postaxial ray) fuse and give rise to the singular cartilage rudiment of the radiale-intermedium (fig. 5a). This also remains the final pattern after ossification.

Another type of fusion occurs at later stages of development, during the onset of pronounced growth of the cartilage rudiments. In this case the well-defined cartilage elements fuse together, probably through atrophy of the perichondrial cells surrounding each element. An example is the incorporation of the tibiale and fibulare of the bird ankle into the tibia rudiment, resulting in the formation of the tibiotarsus typical for birds (fig. 5b).

A third mechanism can produce changes of skeletal patterns at even later stages of ontogeny. These are fusions achieved during the onset of ossification, which marks the next stage of skeletogenesis. Through the encroachment of the ossification process several formerly individual cartilage rudiments can be incorporated into one single bone. This is the case in the tibio-fibular fusion of rodents, and in the formation of the os antebrachii and the os cruris of frogs or the tarsometatarsus of birds (fig. 5c).

Such cases illustrate that, contrary to intuitive assumptions, the phylo-

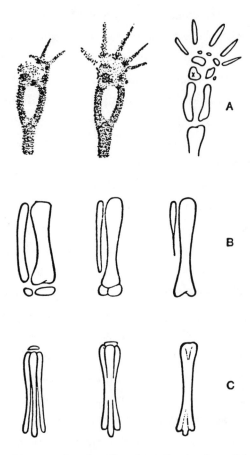

Figure 5. A. Fusion of two carpal condensations forms the singular radiale-intermedium (x) in *Alligator mississippiensis* (after Müller and Alberch 1990). B. Fusion of two tarsal cartilages to the tibia rudiment forms the tibiotarsus in chick ontogeny. C. Fusion of three separate metacarpal rudiments during the ossification phase of the chick tarsometatarsus.

genetic modification of skeletal patterns is not based on a singular ontogenetic mechanism and is not restricted to alterations of the pattern-forming processes, but actually can be realized through a variety of mechanisms that govern rather late stages of ontogeny. As a rule, opportunities for alterations arise together with the takeover of a new mechanism in ontogeny, e.g., when osteogenesis is replacing chondrogenesis. We should, therefore, be able to predict that more opportunities for phylogenetic modifications exist in organs that pass through a large number of ontogenetic stages, involving multiple developmental mechanisms. Another important corollary is that the preexistence of developmental mechanisms which simply spread to other areas, such as ossification incorporating several cartilage elements, holds the potential for rapid morphological transformation, as it does not necessitate the establishment of new developmental mechanisms.

### Plasticity, Integration, and Amplification

Having discussed some properties of developmental systems that promote the generation of morphological novelty, it is also necessary to explore the chances of successful incorporation of such innovations into the organism. The coordination of novel structures with the associated organ systems is obviously a prerequisite for the maintenance of a viable and functionally integrated phenotype, both embryonic and adult. It has repeatedly been pointed out in the past (Raff and Kaufman 1983) that the plasticity, the regulatory capacity, and the genetic-epigenetic feedback loops of developmental systems sufficiently account for coordination at the cellular, tissue, and organ levels of development. Even at very early stages of development a surprising flexibility can be noted (Raff et al., this volume). Not much needs to be added here, except to highlight a few aspects of plasticity in the context of novelty.

While plasticity and regulation are almost trivial to the embryologist, these properties seem to be largely underestimated in their contribution to morphological evolution. This is probably due to the prevalence of gradualistic genome-centered concepts of evolution, as expressed by the frequently encountered argument that major mutational changes would drastically interfere with the harmony of development and would almost certainly be deleterious (Mayr 1976). Such assertions evidently depend on what is considered a major change. If, for example, it is the alteration of a protein that plays an important role in the cell-cell interactions of early embryogenesis, then this

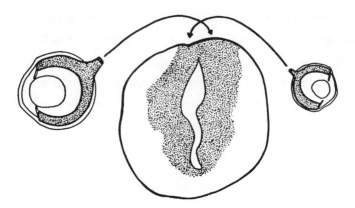

Figure 6.   Transplantation of the eye primordium from the large salamander *Ambystoma tigrinum* results in the development of a large eye (left) in the small-eyed larva of *Ambystoma punctatum*. The increased input of optic neurons triggers an augmentation of cell number in the contralateral optic thalamus. After Twitty 1932.

can be based on a very minor alteration of a single structural gene, but its vast effects are very unlikely to become integrated during the further course of development.   On the other hand, morphological novelty will rarely, if ever, arise from this kind of mutation, and if we consider mutations that affect size, proportions, or rates, then we may deal with larger structural modifications that nevertheless can be easily accommodated by the plasticity of developmental systems.   I will describe some examples based on experimental studies, because many techniques in experimental embryology create ontogenetic alterations comparable to evolutionary novelty.   Most of these are by far more drastic than any natural innovation could possibly be, and still the systems are able to integrate such abrupt perturbations.

One of many examples to illustrate this is Twitty's famous xenoplastic eye transplant (fig. 6).   Twitty (1932) grafted the eye primordium of a large salamander (*Ambystoma tigrinum*) orthotopically to the head region of a much smaller salamander (*A. punctatum*) with smaller eyes.   Although the eventual size of the transplant was found to be intrinsically controlled, i.e., the grafted eye grew to a size corresponding to the large donor animal, it induced several regulatory responses in the host heads.   Not only did the cartilaginous orbital capsule grow larger than that of the contralateral control, but the ocular muscles associated with the transplanted eye also increased in number of muscle fibers. Most exciting, however, is the fact that the larger number of retinal ganglion cells in the transplanted bulbs induced changes in the associated optic brain centers. The preoptic nucleus and the gray matter of the optic thalamus underwent changes in cell number and arrangement, and a 10 to 20% increase of tectal neurons appeared in the contralateral projection areas of the midbrain. Cytologically this is probably based on the maintenance of an initial excess of tectal neurons.

This, and similar experiments using olfactory placodes (Burr 1930), shows how a highly discoordinated modification is easily integrated through the plasticity of developmental mechanisms, a property that was termed "ontogenetic buffering" (Waddington 1957; Katz et al. 1981). It is particularly striking how well functionally interdependent structures are also epigenetically coupled in ontogeny, a rule that especially holds for the interactive regulation of skeletal, muscular, and neuronal development. Experimental changes in the skeletal system set off coordinated changes in the muscle system (Scott 1957) that can even parallel the patterns of interspecific variation of these muscles (Müller 1986). The patterns of innervation, in turn, can respond to changes in the muscle system through mechanisms of target finding (Wilson et al. 1988) and inductive matching of the number of muscle fibers with the numbers of motoneurons (Katz and Lasek 1978). Changes in the number of peripheral motoneurons can be accommodated in the central nervous system by neuronal plasticity at the level of the spinal cord (Detwiler 1920; Hollyday and Hamburger 1976).

In the case of vertebrate limbs, where most of these mechanisms were studied, it is difficult to imagine any kind of novelty that could not be readily and automatically integrated by the epigenetic cascade, and the physiological viability of a new structure is always guaranteed by the most plastic system of all, the vascular system. By this I do not imply that any mutational event will necessarily become integrated into a viable phenotype, but novelties of a magnitude to be generated by mechanisms discussed in the previous sections can be easily coordinated with the associated systems. The high correlation between functional and epigenetic coupling of organ systems is even considered to be of adaptive evolutionary significance (Wagner 1984; Bürger 1986; Müller 1989) and may rapidly generate integrated morphological complexes that drastically increase the adaptive versatility of a group (Liem 1974).

In addition to its regulatory capacities, the epigenetic nature of development also accounts for the fact that relatively small initial changes in morphogenesis, a small shift in epithelial invagination, a small new contact area of formerly separate tissues, a slight change in the onset or termination of ossification, etc., can be magnified into a prominent phenotypic effect during the further course of development – a phenomenon we may call amplification. Such effects of epigenetic amplification, together with the threshold effects, may contribute fundamentally to the often-proposed rapidity in the origination of novelty, which underlies punctuational paleontological scenarios (Gould and Eldredge 1977; Carroll 1988).

## Conclusion

Ontogeny, the mediating system between the genome and the phenotype, is in evolutionary theory treated as a kind of black box. In a scientific climate that mainly promotes the study of interactions between environment and the genetics of populations, embryonic mechanisms are often taken as mere translators of genome evolution into phenotypes. The insufficiencies of this adaptationist program have been pointed out in numerous accounts (Gould and Lewontin 1978; Ho and Saunders 1979; Alberch 1982; John and Miklos 1988). One of its main flaws is that population genetics cannot be predictive of organismic structures. In contrast to such traditionally externalist approaches to evolution the question of novelty is a genuinely internalist one, a question about the specificity of morphological design, and about the organismic determinants generating this specificity. (For recent discussions of externalist versus internalist programs see Lauder 1982; Wake and Larson 1987). The increased interest in the determinants of evolutionary innovation, as documented by the present volume, reflects a shift of attention from external to internal factors, based on the growing sense that this neglected area may hold the key to some macroevolutionary phenomena such as trends, stasis, parallels, discontinuity, novelty, etc. By now a fairly broad conceptual foundation for the role of internal evolutionary factors, in part based on experimental evidence, has been established (e.g., Gould 1977; Bonner 1982; Alberch et al. 1979; Alberch 1980; Raff and Kaufman 1983; Riedl 1978; Wagner 1986), and can be used as a heuristic platform for the expansion of the field.

The ideas developed in this essay attempt to fill the gap that exists between formal concepts and mechanistic explanations concerning the relationship between development and evolution. The focus, however, is on novelty, refraining from a detailed discussion of other aspects of morphological transformation. It is also recognized that the separation and sequential discussion of the various mechanisms is artificial and does not take full account of the interrelated nature of these processes. The following tentative concept emerges from the observations made in the previous sections.

It is not disputed that the primary causes of evolution are Darwinian, but it seems that at the developmental level genomic changes will essentially influence quantitative developmental parameters and will not cause qualitative novelty per se. This quantitative and possibly gradual modification of the rates and timing of developmental events can eventually bring the affected systems to their limits. Novelties arise at the transgression of threshold points immanent to developmental systems, where sequences of developmental interaction are disrupted or new interactions become established, and the resulting structure will depend on the reaction norms of the system at that point. This may directly translate into the adult phenotype or may initially produce a transitory structure in ontogeny not immediately expressed in the adult. Such transitory structures can later become expressed in the adult stages of descendants through processes of heterochrony or they can become further modified and provide a developmental basis for other morphological innovations. Particular opportunities for the modification of developmental patterns exist during the transition phases of changing mechanisms in the stepwise formation of organs. Generally it can be expected that new structures arising are easily integrated into the organism through epigenetic adjustments of the associated systems.

The argument presented contains two main principles. One is that tissue reactions to changing developmental conditions during evolution can produce intermediate structures in ontogeny. These caenogenetic structures are, therefore, not direct consequences of external selection processes but are byproducts of epigenesis. In many instances morphological novelty in evolution may be based on such caenogenetic structures, an example of which is given by Müller and Streicher (1989) for the evolution of the bird hindlimb. The important evolutionary role assigned to caenogenesis in the classic works of Garstang (1922), de Beer (1958), and Rensch (1959) has probably been largely misunderstood, because caenogenesis was mainly viewed as a deviation from recapitulatory processes and was rejected along with the biogenetic law. The second aspect of caenogenetics, "reaction to embryonic conditions," is the one taken up here, emphasizing that caenogenetic structures themselves can be epigenetic side effects of changing rates in developmental processes.

The second principle and centerpiece of the argument is that the threshold properties of developmental systems can account for the origin of discontinuity in the variation of characters. If experimental alterations of development produce discrete classes of effects, there is no logical reason why genetic alteration of the same processes must have continuous effects. Developmental thresholds may, therefore, be seen as a key component of the morphogenetic interface between genetic variation and morphological novelty. Threshold concepts are not new in evolutionary theory, but they have mainly been restricted to quantitative genetics (e.g., Grüneberg 1952; Falconer 1981) or to statistical studies based on phenotypic characters (Lande 1977). Several formal models of development are also implicitly or explicitly based on threshold notions (Turing 1952; Crick 1970; Murray 1981), but no attempt has been made so far to integrate evolutionary and developmental concepts of thresholds. This will depend on a broader empirical understanding of threshold properties in ontogenetic systems, including in particular those of developmental cell mechanics.

If morphological novelties are initially epigenetic by-products which arise as a consequence of threshold properties in development, it follows further that it is not necessary to evoke new genes for their origin, as had been proposed on previous occasions (Stebbins 1968). Rather, we may find at the genome level an epigenetically induced, modified activation of existing genes. This does not exclude the possibility of later genetic assimilation of the new character (Waddington 1957; Van Valen 1974; Riedl 1978) and its exposure to natural selection, but genetic mechanisms will not of necessity have to be held responsible as initiating causal agents. In addition, it is noteworthy that the three properties of development discussed – threshold phenomena, intermediate structures, and sequential transition of mechanisms – share the capacity to produce discontinuity within brief periods of time. This inherent potential for relatively rapid morphological transformation could to some extent account for the apparent lack of intermediate forms in the fossil record. This makes the threshold concept compatible with punctuational theories of evolution that advocate periods of rapid morphological transition (Gould and Eldredge 1977). If the hypothesis holds that the main causality for the origin of novelty lies in epigenesis, then it will be necessary for a next synthesis in evolutionary theory to incorporate a theory of development.

In summary, the proposed hypothesis for the origin of morphological novelty consists of the following principal points: (1) Novelties are defined as qualitative morphological changes with a discontinuous deviation from the ancestral state. (2) The majority of novelties arise as secondary by-products of epigenesis that appear when quantitative modifications of developmental processes reach a threshold of the affected system. (3) The causality for the origin of novel structures lies not within the genome but in epigenesis. (4) The specificity of morphological innovations depends on the reaction norms of developmental systems at their limits. (5) Intermediate structures and sequential switching of mechanisms provide opportunities for novelties to arise. (6) The plasticity prevailing at all levels of developmental systems facilitates the epigenetic integration of novelties. (7) The mechanisms named share a potential for rapid morphological transformation which may underlie several discontinuous phenomena of the evolutionary record.

### Acknowledgments

This paper was written in the fall of 1988 while I was on a sabbatical leave at the Department of Organismic and Evolutionary Biology at Harvard University. I wish to express my gratitude for the cordial reception and the stimulating discussions of ideas expressed herein to Pere Alberch, Ann Burke, Emily Gale, Ed Gilland, Greg Mayer, John Reiss, Chris Rose, and Ernest Williams. The comments of Ann Burke, Emily Gale, Brian Hall, and Tom Miyake on earlier versions of this manuscript were particularly helpful. Kathy Brown-Wing drew the illustrations. L. J. Cook and the Cantab crew have contributed as a continuous source of joy and energy. This work was supported by grant #J0303M of the Fonds zur Förderung der wissenschaftlichen Forschung. Initial work on which this paper is based was supported by NSERC International Exchange Award #39483.

# References

Alberch, P. 1980. Ontogenesis and morphological diversification. *American Zoologist* 20: 653-67.

Alberch, P. 1982. Developmental constraints in evolutionary processes, pp. 313-32. In *Evolution and Development*, ed. J. T. Bonner. Berlin, Heidelberg, New York: Springer-Verlag.

Alberch, P. 1985. Developmental constraints: Why St. Bernards often have an extra digit and Poodles never do. *American Naturalist* 126: 430-33.

Alberch, P., and E. A. Gale. 1983. Size dependence during the development of the amphibian foot. Colchizine-induced digital loss and reduction. *Journal of Embryology and Experimental Morphology* 76: 177-97.

Alberch, P., and E. A. Gale. 1985. A developmental analysis of an evolutionary trend: Digital reduction in amphibians. *Evolution* 39: 8-23.

Alberch, P., S. J. Gould, G. F. Oster, and D. B. Wake. 1979. Size and shape in ontogeny and phylogeny. *Paleobiology* 5(3): 296-317.

Anthony, J., and J. Guibe. 1952. Les affinités anatomiques de *Bolyeria* et de *Casarea*. *Mémoires de l'Institut Scientifique de Madagascar (Série A)* 7: 189-201.

Bertalanffy, L. 1952. *The Problem of Life*. New York: Harper.

Bonner, J. T., ed. 1982. *Evolution and Development*. Berlin, Heidelberg, New York: Springer-Verlag.

Bretscher, A. 1947. Reduktion der Zehenzahl bei *Xenopus*-larven nach lokaler Colchicinbehandlung. *Revue Suisse de Zoologie* 54: 273-79.

Bretscher, A. 1949. Die Hinterbeinentwicklung von *Xenopus laevis* Daud. und ihre Beeinflussung durch Colchicin. *Revue Suisse de Zoologie* 56: 33-96.

Bretscher, A., and P. Tschumi. 1951. Gestufte Reduktion von chemisch behandelten *Xenopus* Beinen. *Revue Suisse de Zoologie* 58: 391-98.

Brylski, P., and B. K. Hall. 1988a. Epithelial behaviors and threshold effects in the development and evolution of internal and external cheek pouches in rodents. *Zeitschrift für Zoologische Systematik und Evolutionsforschung* 26: 144-54.

Brylski, P., and B. K. Hall. 1988b. Ontogeny of a macroevolutionary phenotype: The external cheek pouches of geomyoid rodents. *Evolution* 42: 391-94.

Bürger, R. 1986. Constraints for the evolution of functionally coupled characters: A nonlinear analysis of a phenotypic model. *Evolution* 40(1): 182-93.

Burke, A. C. 1989. The development of the turtle carapace: Implications for the evolution of a novel Bauplan. *Journal of Morphology* 199: 363-78.

Burr, H. S. 1930. Hyperplasia in the brain of *Ambystoma*. *Journal of Experimental Zoology* 55: 171-91.

Carroll, R. L. 1988. *Vertebrate Paleontology and Evolution.* New York: W. H. Freeman.

Crick, F. 1970. Diffusion in morphogenesis. *Nature* 225: 420-22.

Dalcq, A. 1949. L'apport de l'embryologie causale au problème de l'évolution. *Portugaliae Acta Biologica (Serie A)*: 367-400.

Davis, D. D. 1964. *The Giant Panda: A Morphological Study of Evolutionary Mechanisms. Fieldiana: Zoology Memoirs.* Vol. 3.

de Beer, G. R. 1938. Embryology and evolution, pp. 57-78. In *Evolution: Essays Presented to E. S. Goodrich*, ed. G. R. de Beer. Oxford: Oxford University Press.

de Beer, G. R. 1958. *Embryos and Ancestors.* Oxford: Clarendon Press.

Detwiler, S. R. 1920. On the hyperplasia of nerve centers resulting from excessive peripheral loading. *Proceedings of the National Academy of Sciences* 6: 96-101.

Dobzhansky, T., F. J. Ayala, G. L. Stebbins, and J. W. Valentine. 1977. *Evolution.* San Francisco: W. H. Freeman.

Ede, D. A., and O. P. Flint. 1972. Patterns of cell division, cell death and chondrogenesis in cultured aggregates of normal and *talpid*³ mutant chick limb mesenchyme cells. *Journal of Embryology and Experimental Morphology* 27: 245-60.

Ede, D. A., O. P. Flint, O. K. Wilby, and P. Colquhoun. 1977. The development of the pre-cartilage condensations in the limb bud mesenchyme *in vivo* and *in vitro*, pp. 161-79. In *Vertebrate Limb and Somite Morphogenesis*, ed. D. A. Ede, J. R. Hinchliffe and M. Balls. Cambridge: Cambridge University Press.

Falconer, D. S. 1981. *Introduction to Quantitative Genetics.* London: Longman.

Feduccia, A. 1980. *The Age of Birds.* Cambridge: Harvard University Press.

Frazzetta, T. H. 1970. From hopeful monsters to bolyerine snakes? *American Naturalist* 104: 55-72.

Gans, C. 1975. Tetrapod limblessness: Evolution and functional corollaries. *American Zoologist* 15: 455-67.

Garstang, W. 1922. The theory of recapitulation: A critical restatement of the biogenetic law. *Journal of the Linnean Society. Zoology* 35: 81-101.

Goldschmidt, R. 1940. *The Material Basis of Evolution.* New Haven: Yale University Press.

Gould, S. J. 1977. *Ontogeny and Phylogeny.* Cambridge: Harvard University Press.

Gould, S. J. 1980. *The Panda's Thumb.* New York: Penguin Books.

Gould, S. J. 1982. Change in developmental timing as a mechanism of macroevolution, pp. 333-46. In *Evolution and Development*, ed. J. T. Bonner. Berlin, Heidelberg, New York: Springer-Verlag.

Gould, S. J., and N. Eldredge. 1977. Punctuated equilibria: Tempo and mode of evolution reconsidered. *Paleobiology* 3: 115-51.

Gould, S. J., and R. C. Lewontin. 1978. The spandrels of San Marco and the Panglossian paradigm: A critique of the adaptationist programme. *Proceedings of the Royal Society of London* 205: 581-98.

Grüneberg, H. 1952. Genetical studies on the skeleton of the mouse. *Journal of Genetics* 51: 95-114.

Haeckel, E. 1866. *Generelle Morphologie der Organismen.* 2 Vols. Berlin: Reimer.

Hall, B. K. 1970. Cellular differentiation in skeletal tissues. *Biological Reviews (Cambridge Philosophical Society)* 45: 455-84.

Hall, B. K. 1975. Evolutionary consequences of skeletal differentiation. *American Zoologist* 15: 329-50

Hall, B. K. 1978. *Developmental and Cellular Skeletal Biology.* New York: Academic Press.

Hall, B. K. 1979. Selective proliferation and accumulation of chondroprogenitor cells as the mode of action of biomechanical factors during secondary chondrogenesis. *Teratology* 20: 81-92.

Hall, B. K. 1983. Epigenetic control in development and evolution, pp. 353-79. In *Development and Evolution,* ed. B. C. Goodwin, N. Holder, and C. G. Wylie. Cambridge: Cambridge University Press.

Hall, B. K. 1984. Developmental processes underlying heterochrony as an evolutionary mechanism. *Canadian Journal of Zoology* 62: 1-7.

Hall, B. K. 1986. The role of movement and tissue interactions in the development and growth of bone and secondary cartilage in the clavicle of the embryonic chick. *Journal of Embryology and Experimental Morphology* 93: 133-52.

Haluska, F., and P. Alberch. 1983. The cranial development of *Elephe obsoleta* (Ophidia, Colubridae). *Journal of Morphology* 178: 37-55.

Hamburger, V. 1980. Embryology and the modern synthesis in evolutionary theory, pp. 97-112. In *The Evolutionary Synthesis,* ed. E. Mayr and W. B. Provine. Cambridge: Harvard University Press.

Hamburger, V., M. Balaban, R. Oppenheim, and E. Wenger. 1965. Periodic motility of normal and spinal chick embryos between 8 and 17 days of incubation. *Journal of Experimental Zoology* 159: 1-14.

Hanken, J. 1985. Morphological novelty in the limb skeleton accompanies miniaturization in salamanders. *Science* 229: 871-74.

Hinchliffe, J. R., and D. R. Johnson. 1980. *The Development of the Vertebrate Limb.* Oxford: Clarendon Press.

Ho, M. W., and P. T. Saunders. 1979. Beyond neodarwinism – An epigenetic approach to evolution. *Journal of Theoretical Biology* 78: 573-91.

Hollyday, M., and V. Hamburger. 1976. Reduction of the naturally occurring motor neuron loss by enlargement of the periphery. *Journal of Comparative Neurology* 170: 311-20.

John, B., and G. Miklos. 1988. *The Eukaryote Genome in Development and Evolution*. London: Allen & Unwin.

Katz, M. J., and R. J. Lasek. 1978. Evolution of the nervous system: Role of ontogenetic mechanisms in the evolution of matching populations. *Proceedings of the National Academy of Sciences* 75: 1349-52.

Katz, M. J., R. Lasek, and I. R. Kaiserman-Abramof. 1981. Ontophyletics of the nervous system: Eyeless mutants illustrate how ontogenetic buffer mechanisms channel evolution. *Proceedings of the National Academy of Sciences* 78: 397-401.

Lande, R. 1977. Evolutionary mechanisms of limb loss in tetrapods. *Evolution* 32: 73-92.

Larson, A., E. M. Prager and A. C. Wilson. 1984. Chromosome evolution, speciation and morphological change in vertebrates: The role of social behavior. *Chromosomes Today* 8: 215-28.

Lauder, G. V. 1982. Historical biology and the problem of design. *Journal of Theoretical Biology* 97: 57-67.

Liem, K. F. 1974. Evolutionary strategies and morphological innovations: Cichlid pharyngeal jaws. *Systematic Zoology* 22: 425-41.

Long, C. A. 1976. Evolution of mammalian cheek pouches and a possibly discontinuous origin of a higher taxon (Geomyoidea). *American Naturalist* 110: 1093-1111.

Maynard Smith, J., ed. 1982. *Evolution Now: A Century After Darwin*. San Francisco: W. H. Freeman.

Maynard Smith, J., R. Burian, S. Kauffman, P. Alberch, J. Campbell, B. Goodwin, R. Lande, D. Raup, and L. Wolpert. 1985. Developmental constraints and evolution. *Quarterly Review of Biology* 60: 265-87.

Mayr, E. 1976. *Evolution and the Diversity of Life*. Cambridge: Harvard University Press.

Mayr, E. 1980. Some thoughts on the history of the evolutionary synthesis, pp. 1-48. In *The Evolutionary Synthesis*, ed. E. Mayr and W. B. Provine. Cambridge: Harvard University Press.

Mayr, E. 1982. *The Growth of Biological Thought*. Cambridge: Harvard University Press.

McNamara, K. J. 1982. Heterochrony and phylogenetic trends. *Paleobiology* 8(2): 130-42.

Merrilees, M. J., and M. H. Flint. 1980. Ultrastructural study of tension and pressure zones in a rabbit flexor tendon. *American Journal of Anatomy* 157: 87-106.

Müller, G. B. 1986. Effects of skeletal change on muscle pattern formation, pp. 91-108. In *Development and Regeneration of the Skeletal Muscles*, ed. B. Christ and R. Cihak (*Bibliotheca Anatomica* 29). Basel: Karger.

Müller, G. B. 1989. Ancestral patterns in bird limb development: A new look at Hampe's experiment. *Journal of Evolutionary Biology* 2: 31-47.

Müller, G. B., and P. Alberch. 1990. Ontogeny of the limb skeleton in *Alligator mississippiensis*: Developmental invariance and change in archosaur limbs. *Journal of Morphology* 203: 1-14.

Müller, G. B., and J. Streicher. 1989. Ontogeny of the syndesmosis tibiofibularis and the evolution of the bird hindlimb: A caenogenetic feature triggers phenotypic novelty. *Anatomy and Embryology* 179: 327-39.

Murray, J. D. 1981. A pre-pattern formation mechanism for animal coat markings. *Journal of Theoretical Biology* 88: 161-99.

Newman, S. A. 1977. Lineage and pattern in the developing wing bud, pp. 181-97. In *Vertebrate Limb and Somite Morphogenesis*, ed. D. A. Ede, J. R. Hinchliffe and M. Balls. Cambridge: Cambridge University Press.

Oster, G. F., N. Shubin, J. D. Murray, and P. Alberch. 1988. Evolution and morphogenetic rules: The shape of the vertebrate limb in ontogeny and phylogeny. *Evolution* 42(5): 862-84.

Ostrom, J. H. 1976. *Archaeopteryx* and the origin of birds. *Biological Journal of the Linnean Society* 8(2): 91-182.

Raff, R. A., and T. C. Kaufman. 1983. *Embryos, Genes, and Evolution*. New York: Macmillan.

Raynaud, A. 1985. Development of limbs and embryonic limb reduction, pp. 59-148. In *Biology of the Reptilia*, ed. C. Gans and F. Billet. Vol. 15 (B). New York: John Wiley.

Rensch, B. 1948. Histological changes correlated with evolutionary changes of body size. *Evolution* 2: 218-30.

Rensch, B. 1959. *Evolution Above the Species Level*. New York: Columbia University Press.

Riedl, R. 1978. *Order in Living Organisms*. Chichester: John Wiley.

Schmalhausen, I. I. 1949. *Factors of Evolution*. Philadelphia: Blakiston.

Scott, J. H. 1957. Muscle growth and function in relation to skeletal morphology. *American Journal of Physical Anthropology* 15: 197-234.

Severtzoff, A. N. 1931. *Morphologische Gesetzmässigkeiten der Evolution*. Jena: Gustav Fischer.

Shubin, N. H., and P. Alberch. 1986. A morphogenetic approach to the origin and basic organization of the tetrapod limb. *Evolutionary Biology* 20: 319-87.

Sinervo, B., and L. R. McEdward. 1988. Developmental consequences of an evolutionary change in egg size: An experimental test. *Evolution* 42: 885-99.

Sober, E. 1986. *Conceptual Issues in Evolutionary Biology*. Cambridge: Massachusetts Institute of Technology Press.

Solursh, M. 1984. Ectoderm as a determinant of early tissue pattern in the limb bud. *Cell Differentiation* 15: 17-24.

Stebbins, G. L. 1968. Integration of development and evolutionary progress, pp. 17-36. In *Population Biology and Evolution,* ed. R. Lewontin. Syracuse, NY: Syracuse University Press.

Turing, A. 1952. The chemical basis of morphogenesis. *Philosophical Transactions of the Royal Society of London (B)* 237: 37-72.

Twitty, V. C. 1932. Influence of the eye on the growth of its associated structures, studied by means of heteroplastic transplantation. *Journal of Experimental Zoology* 61: 333-74.

Van Valen, L. 1974. A natural model for the origin of some higher taxa. *Journal of Herpetology* 8: 109-21.

Waddington, C. H. 1941. Evolution of developmental systems. *Nature* 147: 108-10.

Waddington, C. H. 1957. *The Strategy of the Genes.* London: Allen & Unwin.

Waddington, C. H. 1962. *New Patterns in Genetics and Development.* New York: Columbia University Press.

Wagner, G. P. 1984. Coevolution of functionally constrained characters: Prerequisites for adaptive versatility. *BioSystems* 17: 51-55.

Wagner, G. P. 1986. The systems approach: An interface between development and population genetic aspects of evolution, pp. 148-65. In *Patterns and Processes in the History of Life,* ed. D. M. Raup and D. Jablonski. Berlin, Heidelberg, New York: Springer-Verlag.

Wake, D. B. 1981. The application of allozyme evidence to problems in evolutionary morphology, pp. 257-70. In *Evolution Today,* ed. G. Scudder and J. Revel. Pittsburgh: Carnegie-Mellon University.

Wake, D. B., and A. Larson. 1987. Multidimensional analysis of an evolutionary lineage. *Science* 238: 42-48.

Wilson, A. C., S. S. Carlson, and T. J. White. 1977. Biochemical evolution. *Annual Review of Biochemistry* 46: 573-639.

Wilson, S., M. Jesani, and N. Holder. 1988. Reformation of specific neuromuscular connections during axolotl limb regeneration: Evidence that the first contacts are correct. *Development* 103: 365-77.

Wolff, E., and M. Kieny. 1962. Mise en évidence par l'irradiation aux rayons X d'un phénomène de compétition entre les ébauches du tibia et du péroné chez l'embryon de poulet. *Developmental Biology* 4: 197-213.

Wolpert, L., C. Tickle, and M. Sampford. 1979. The effect of cell killing by X-irradiation on pattern formation in the chick limb. *Journal of Embryology and Experimental Morphology* 50: 175-98.

# MORPHOLOGY AND PHYSIOLOGY

# The Evolution of Morphological Variation Patterns

*James M. Cheverud*

Morphological evolutionary innovations can arise by the rearrangement of morphological elements to produce previously unknown combinations of characters, or previously unknown combinations of the subcharacters we designate as individual traits. The possibilities for new trait combinations in evolution will be limited by the patterns of heritable variation for these characters. We would tend to recognize a character as a true innovation when it is inconsistent with the underlying pattern of heritable variation within a species. In this situation the character combination will not appear as an extension of preexisting patterns of variation but rather as a more fundamental reorganization of character complexes.

The pattern of heritable variation for a trait set is measured by the genetic covariance or correlation matrix for the characters involved. Genetic correlations measure the extent to which two traits are coinherited due to pleiotropy, when one gene affects more than one character, and to linkage disequilibrium, when two different genes each affecting a different character are inherited together (Falconer 1981). The pattern of genetic correlation plays a crucial role in determining the rate and direction of evolution in response to both selection and random drift (Lande 1979). A character set's response to selection is given by

$$\Delta z = G\beta \tag{1},$$

where $\Delta z$ is an $n \times 1$ vector describing the change in means for each of $n$ characters, $G$ is the $n \times n$ genetic variance/covariance matrix describing the

pattern of heritable variation, and ß is the n x 1 vector describing direct selection on each character. The evolutionary trajectory (Δz) will not follow
the optimal path of maximum increase in fitness (ß) but will be deflected
from that path if it is inconsistent with the pattern of genetic variation (G).
Thus the pattern of genetic variation can constrain the character combinations that can be expected to evolve in a reasonable period of time. The loss
of evolutionary freedom due to patterned genetic variation is an expected or
even necessary outcome of adaptive evolution (Riedl 1978; Wagner 1988).

From the relationship given above, it seems that heritable variation
patterns can limit the scope of evolutionary innovations by restricting the
range of likely character combinations. The extent of this limitation will depend, in part, on the evolutionary lability of the variation patterns themselves. If patterns of variation evolve freely and quickly, they may not re-
strain innovations over even the short term. However, if variation patterns
are stable over long periods of evolutionary time, they will act more strongly
in limiting innovation.

Lande (1979, 1980) has argued that patterns of heritable variation
should remain relatively stable over time based on a Gaussian approximation
to the infinite alleles model. Variation patterns should not change much
with weak directional selection but would evolve in response to changes in
the pattern of stabilizing selection and changes in mutational effects on characters. In this theoretical model, the pattern of genetic correlation, and thus
the pattern of constraint on innovation, will come to match roughly the pattern of functional interrelationship of the characters (Lande 1980; Cheverud
1982, 1984, 1988a, 1988b). Wagner (n.d.) has indicated that selection on
mutational patterns is likely to be weak at best, so that stability of mutational
patterns should be the rule. Likewise, since much stabilizing selection occurs
due to the effects of a character's interaction with its internal organismal environment (other characters) (Riedl 1978; Cheverud 1984, 1988a), patterns of
stabilizing selection are likely to remain constant over long periods of evolutionary time. Thus, from this perspective, stability of correlation patterns

should be the general rule.

Turelli and coworkers (Turelli 1988a; Barton and Turelli 1987) have developed models based on a "house of cards" approximation to the infinite alleles model, which suggest that patterns of genetic variation and covariation should be quite labile in evolutionary terms. Genetic correlations should change in unpredictable ways due to directional selection and changes in the environment (Turelli 1988b). If genetic correlations are quite labile, they will not be very effective in limiting morphological innovations whereas, if they are stable over long periods of time, evolutionary innovations will be less likely to occur. Heritable variation patterns are seen here as permissive, allowing or disallowing the production of specific innovations.

Since the theoretical models come to contradictory conclusions, the question of heritable variation pattern stability becomes chiefly an empirical one. I will present a comparative analysis of phenotypic variation patterns for facial characters in seven species of papionin primates, including macaques, baboons, and mangabeys, in order to determine the relative stability of variation patterns within this group (Cheverud, in press).

Phenotypic rather than genetic variation patterns will be analyzed because genetic data are not available on most papionin species. It is quite difficult to obtain reliable genetic correlation estimates due to the large number of related individuals required for accurate measurement. In a comparative analysis of the relationship between genetic and phenotypic correlation structures based on a wide variety of studies of morphological traits, I (1988b) found that genetic and phenotypic correlation patterns were broadly similar, the average matrix correlation between genetic and phenotypic correlation matrices being 0.57. However, the best estimated genetic correlation matrices, those derived from large samples of related individuals, were very highly correlated with their phenotypic counterparts (average matrix correlation = 0.81) indicating that the relatively large sampling error involved in estimating genetic correlations is responsible for much of the observed difference between genetic and phenotypic correlations. Also, it is difficult to imagine the scenario in which similarity of phenotypic correlation structure among species disguises a diversity of genetic correlation for morphological characters. Thus, it seems reasonable, and eminently practical, to carry out comparisons of correlation structure at the phenotypic level, even though the theoretical issues concern only the genetic portions of the variation. Comparisons at the genetic level (Kohn and Atchley 1988; Lofsvold 1986; Atchley et al. 1981; Arnold 1981) must be made when possible, but are likely to remain rare and restricted to only a few species.

### Finite Element Scaling Measures

The characters to be analyzed are morphologically local size differences relative to an average infant female rhesus macaque. These local size differences are obtained from the strain tensors of a finite element scaling analysis (Cheverud and Richtsmeier 1986). This form of measurement is particularly suited to studies of morphological integration and variance patterns because it allows for the relatively independent *measurement* of morphological characters. Many standard craniometric techniques and measurement sets contain traits which are necessarily correlated because of their geometrical arrangement, so that similarities in correlation structures among species may be due to similarities in the geometry of the measurements chosen by the investigator rather than similarities of biological processes. This is an undesirable characteristic of many morphometric data sets (Cheverud and Richtsmeier 1986). Indeed, if two linear dimensions share a point in common, the spurious geometrical correlation between them will be $(1/2) \cos \theta$. Finite element scaling allows one to control for this spurious correlation between measurements.

In a finite element scaling analysis, the skull is divided into several hexahedral, cubelike finite elements. The vertices, or corners, of these elements in the various skulls are homologous landmarks. The elements are used to model the complex morphology of the skull. The accuracy of the model depends on the size of the elements – the smaller the elements, the more accurate the model – and on the complexity of the differences between the compared forms – the more homogeneous the contrast between forms across each element, the more accurate the model. The elements of a reference skull, in this case an average infant rhesus monkey (*Macaca mulatta*), are deformed into each individual target skull, in this case an adult from one of the seven papionin species, and the extent of the deformation is measured by the morphometric strain necessary to accomplish it.

With the nonhomogeneous strain hexahedral elements employed here, strain is measured separately at each of the landmarks used to define the corners of the elements allowing a localized measure of size at each point. These measurements are independent from element to element but are not entirely independent from point to point within an element. Simulation studies show that landmarks within a finite element that are connected by an edge are functionally related to one another by the mathematics of the method and are morphometrically correlated at a level of 0.30. This correlation is generated by the assumption that the edge connecting two points maintains its integrity across the comparison. The assumption and its resultant correlation are biologically most reasonable when the landmarks are connected by an osseous edge. This functional correlation due to the structure of the finite element model is very specific and can be easily controlled for in comparative studies of variation. For further information on finite element scaling methods see Cheverud and Richtsmeier (1986), Richtsmeier and Cheverud (1986) and Lewis et al. (1980).

In this study, two elements are defined using twelve landmarks (table 1) representing the right side of the lower and upper face. The lower facial element is bounded by oral landmarks along the alveolus (points 1, 2, and 4) and palate (pt. 6) and landmarks describing the upper boundary of the nasal cavity (pts. 7, 8, 10, and 12). The superior boundary of the upper facial element is composed of points along the upper edge of the orbits (pts.

Table 1.

Osteological landmarks on the right side of the face used in finite element analysis. Point numbers refer to those used in Cheverud and Richtsmeier (1986). The lower facial element is composed of points 8, 10, 12, 7, 2, 4, 6, and 1, while the upper facial element is composed of points 16, 18, 41, 13, 8, 10, 12, and 7.

| | |
|---|---|
| 1. intradentale superior | |
| 2. premaxilla-maxilla junction at alveolus | 12. inferior vomer-sphenoid junction |
| 4. posterior maxilla at alveolus | 13. nasion |
| 6. posterior nasal spine | 16. fronto-zygomatic junction |
| 7. inferior internasal junction | 18. fronto-spheno-zygomatic junction |
| 8. zygomaxillare superior | 41. point midway between the posterior |
| 10. superior border of the pterygo-palatine | aspects of the orbits |
| fossa | |

13, 16, 18, and 41). The three dimensional coordinates of the landmarks were obtained using a diagraph. These coordinate values were used to define finite elements for the finite element scaling procedure which then produced measures of size local to each landmark for use in morphometric analysis.

## Samples

The seven species studied here are from the tribe Papionini, including four macaque species: *Macaca mulatta* (RHE, N=71), *M. fascicularis* (FAS, N=80), *M. assamensis* (ASS, N=11), *M. nemestrina* (NEM, N=12); two baboon species: *Papio anubis* (ANU, N=16), *P. cynocephalus* (CYN, N=18); and one mangabey species: *Cercocebus albiginea* (ALB, N=42). The RHE, CYN, and ALB samples are from single localities (Cheverud, in press) while the others were collected across their species range. The taxonomic status of the CYN sample is in doubt, as the collection site is near the species boundary between *P. cynocephalus* and *P. anubis* and thus may contain hybrids or *P. anubis* individuals. There is also some controversy over whether these two baboon taxa should be separated at the specific or subspecific level (Szalay and Delson 1979).

## Methods of Comparison

All animals in the sample were fully adult, as defined by completion of dental eruption. The correlations among the twelve characters were obtained from the residual correlation matrices of a MANOVA with sex as the factor and the twelve characters as dependent variables. Thus, all correlations are pooled across sexes. The correlation matrices obtained will be compared both in terms of magnitudes and patterns of correlation using randomization methods (Cheverud et al., in press). The overall level of correlation is measured by the variance of the eigenvalues [V(L)] obtained through spectral decomposition of the correlation matrix (Wagner 1984; Cheverud et al., in press). As this metric has no tractable probability distribution, confidence intervals were obtained by a bootstrap resampling strategy (Cheverud et al., in press). Raw eigenvalue variances are also corrected for the upward bias introduced by small sample sizes by subtracting the expected eigenvalue variance given no correlations between characters in the true population from the observed eigenvalue variance (Wagner 1984; Cheverud et al., in press).

Similarities in the structure of correlation matrices will be assessed using a matrix correlation (Sneath and Sokal 1973) which is a Pearson product moment correlation between the paired elements of the two matrices. Since this statistic also has no known probability distribution, a Mantel's randomization test will be performed (Cheverud et al., in press). This test is sensitive to error in the estimates of individual correlation values which are assumed to be measured without error. To the extent that the correlation values are estimated with error, the matrix correlation between two observed matrices will be reduced, perhaps below the level of statistical significance. In order to judge the stability of the correlation estimates for each species, a bootstrap procedure was used to obtain a distribution of matrix correlations between the observed matrix and each matrix obtained through resampling. The geometric mean of this distribution is taken as a measure of the stability of the observed matrix with respect to the current study. It is suggested

(Cheverud et al., in press) that the matrix correlation between two observed matrices is limited in magnitude by their stability and that the matrix correlation can be roughly corrected for instability of the observed matrices by dividing the matrix correlation by the square root of the product of the stability measures obtained for each individual observed matrix. The stability of individual observed matrices for the species used is given in Cheverud (in press).

## Comparison of Correlation Magnitudes and Patterns

The overall magnitude of correlation for each of the seven matrices, as measured by the corrected eigenvalue variance, is given in table 2. Values range from 1.14 for RHE to 4.05 for CYN, with 95% confidence intervals from the 2.5 and 97.5 percentiles of the bootstrapped distribution. Each matrix shows statistically significant integration of characters. Of the 21 interspecific comparisons only 3 (14% of the total) are statistically significant. The significant contrasts include RHE with both ALB and CYN and FAS with CYN so that

Table 2.

Overall magnitude of correlation in each observed correlation matrix as measured by the variance in eigenvalues [V(L)]. See text for abbreviations.

| Species | V(L) | 95% Confidence Interval |
|---------|------|-------------------------|
| ALB | 3.12 | 1.91 to 4.42 |
| CYN | 4.05 | 2.65 to 5.52 |
| ANU | 1.60 | 1.00 to 2.31 |
| ASS | 2.29 | 0.92 to 4.32 |
| NEM | 2.20 | 1.23 to 3.57 |
| FAS | 1.74 | 1.27 to 2.26 |
| RHE | 1.14 | 0.76 to 1.66 |

both RHE and CYN figure in two significant comparisons. Given the multiple dependent comparisons, little or no distinction among species is likely in the overall magnitude of correlation.

The correlations among the seven observed correlation matrices (table 3) range from 0.28 to 0.75 with an average of 0.53. All are significantly different from zero at the 5% level. After correction for lack of stability in individual correlation estimates, the matrix correlations range from 0.33 to 0.90 with an average of 0.61. All correlation values below 0.45 involve the NEM sample. This is an especially small sample which showed the lowest reliability in the analysis of matrix stability (Cheverud, in press). Overall, the correlation structures are quite similar across the seven species.

The patterns of correlation exhibited by the seven species, and the similarities among these patterns, are not due to correlations caused by the finite element method. The average matrix correlation between the observed correlations and those resulting from the finite element method is 0.14. Also, when the correlations due to finite element scaling are partialled out of the interspecific comparisons (Cheverud, in press; Dow et al. 1987), there is only a very slight reduction in the interspecific correlation values with all of them remaining statistically significant.

Table 3.

Matrix correlations between observed correlation matrices for the seven papionin species. See text for abbreviations.

|      | ALB  | CYN  | ANU  | ASS  | NEM  | FAS  | RHE  |
|------|------|------|------|------|------|------|------|
| ALB  | 1.00 | 0.48 | 0.63 | 0.75 | 0.51 | 0.62 | 0.47 |
| CYN  |      | 1.00 | 0.54 | 0.58 | 0.32 | 0.51 | 0.62 |
| ANU  |      |      | 1.00 | 0.62 | 0.28 | 0.57 | 0.57 |
| ASS  |      |      |      | 1.00 | 0.32 | 0.61 | 0.50 |
| NEM  |      |      |      |      | 1.00 | 0.59 | 0.44 |
| FAS  |      |      |      |      |      | 1.00 | 0.53 |
| RHE  |      |      |      |      |      |      | 1.00 |

The pattern most evident in all seven correlation matrices is that the point of junction between the frontal, zygomatic, and sphenoid bones on the external surface of the orbit (pt. 18) is only weakly correlated with the rest of the landmarks. This point is near the line of insertion of the anterior portion of the temporalis muscle, and its independent variability may be due to local remodeling in response to muscular forces. All matrices, except that for NEM, display relatively high correlations among the alveolar landmarks (intradentale superior; premaxilla-maxilla junction at the alveolus; posterior maxilla at alveolus). All three of these regions are affected by the development and function of the oral cavity.

The variation in correlation matrices among the seven species was compared to their phylogenetic history in order to determine whether similarity has a historical component. The tribe Papionini is considered monophyletic, having diverged from the rest of the Cercopithecids about 10 million years ago (Delson 1980; Szalay and Delson 1979). The baboons and mangabeys separated from the macaque lineage about 6 million years ago and from each other about 5 million years ago. The two species (or subspecies) of baboon are separated by about 0.5 million years, although introgression has since occurred. *Macaca assamensis* and *M. nemestrina* separated from the other macaques and each other about 1.5 million years ago while *M. mulatta* and *M. fascicularis* separated about 1 million years ago. The times since separation were used to build a matrix of temporal phylogenetic distance, which was then correlated with the matrix of interspecific correlations. A nonsignificant matrix correlation of -0.10 indicates that variation in correlation patterns across species is independent of their evolutionary relationships.


## Discussion


The magnitudes and patterns of phenotypic correlation between facial traits among the Papionini has remained relatively stable over the past 6 million years. Variation among the species is random with respect to phylogeny, indicating the possibility of rather quick, undirected, but limited evolutionary change in correlation pattern within the context of a single, stable pattern of correlation held in common throughout the group. Thus, at this taxonomic level we do not see the rearrangement of patterns of variation which would make it easy for evolutionary innovations to appear. Instead, a common pattern of constraint is found across the tribe. This stability of variation patterns will act to restrict innovation, making true morphological innovations rare in evolution, especially in contrast to the frequency of interspecific morphological differences consistent with patterns of within-species variation.

Perhaps differences in correlation patterns will be observed at higher taxonomic levels, where more striking functional differences occur. From this perspective it would be interesting to compare this data with data obtained from the leaf-eating colobine monkeys which are members of the same primate family (Cercopithecidae) as the macaques, baboons, and mangabeys. Olson and Miller (1958) described differences in the pattern of morphological integration among measures taken from lower molars in *Aotus trivirgatus* and *Hyopsodus* specimens. These two genera are in different orders of the mammals and are thus quite distinct in relation to the comparisons given above. Berg (1960) also found differences in the patterns of morphological integration among plant species. He attributed these differences to ecological factors relating to species pollinated by specialized insects versus those species pollinated by wind or unspecialized insects. The scale of comparison in these two studies is much broader than the one presented here. The high level of comparison needed to identify differences in variation patterns is perhaps an indication of the relatively slow rate of change in variation patterns and evolutionary innovations as compared to speciation.

## Acknowledgments

I thank Donald Sade for access to the Cayo Santiago macaque specimens, Neil Tappen for access to the University of Wisconsin–Milwaukee primate skeletal collection, and Bruce Patterson for access to specimens from the Field Museum of Natural History. I also thank Joan Richtsmeier and Steve

Leigh for their help in collecting data and Lyle Konigsberg for data collection, computer programming, and data analysis. Special thanks go to Gunter Wagner for stimulating discussion of the problems addressed here. This work was generously supported by a basic research grant from the Whitaker Foundation.

## References

Arnold, S. J. 1981. Behavioral variation in natural populations. I. Phenotypic, genetic, and environmental correlations between chemoreceptive responses to prey in the garter snake, *Thamnophis elegans*. *Evolution* 35: 489-509.

Atchley, W. R., J. J. Rutledge and D. E. Cowley. 1981. Genetic components of size and shape. II. Multivariate covariance patterns in the rat and mouse skull. *Evolution* 35: 1037-55.

Barton, N. H., and M. Turelli. 1987. Adaptive landscapes, genetic distance and the evolution of quantitative characters. *Genetical Research, Cambridge* 49: 157-73.

Berg, R. 1960. The ecological significance of correlation pleiades. *Evolution* 14: 171-80.

Cheverud, J. M. 1982. Phenotypic, genetic, and environmental morphological integration in the cranium. *Evolution* 36: 499-516.

Cheverud, J. M. 1984. Quantitative genetics and developmental constraints on evolution by selection. *Journal of Theoretical Biology* 101: 155-71.

Cheverud, J. M. 1988a. The evolution of genetic correlation and developmental constraint, pp. 94-101. In *Population Genetics and Evolution*, ed. G. de Jong. Berlin: Springer-Verlag.

Cheverud, J. M. 1988b. A comparison of genetic and phenotypic correlations. *Evolution* 42: 958-68.

Cheverud, J. M. In press. A comparative analysis of morphological variation patterns in the papionins. *Evolution*.

Cheverud, J. M., and J. T. Richtsmeier. 1986. Finite element scaling applied to sexual dimorphism in rhesus macaque (*Macaca mulatta*) facial growth. *Systematic Zoology* 35: 381-99.

Cheverud, J. M., G. P. Wagner, and M. M. Dow. In press. Methods for comparative analysis of variation patterns. *Systematic Zoology*.

Delson, E. 1980. Fossil macaques, phyletic relationships and a scenario of deployment, pp. 10-30. In *The Macaques: Studies in Ecology, Behavior and Evolution*, ed. D. G. Lindburg. New York: Van Nostrand.

Dow, M. M., J. M. Cheverud, and J. Friedlaender. 1987. Partial correlation of distance matrices in studies of population structure. *American Journal of Physical Anthropology* 72: 343-52.

Falconer, D. S. 1981. *Introduction to Quantitative Genetics*. London: Longman Press.

Kohn, L. A., and W. R. Atchley. 1988. How similar are genetic correlation structures? Data from mice and rats. *Evolution* 42: 467-81.

Lande, R. 1979. Quantitative genetic analysis of multivariate evolution, applied to brain:body size allometry. *Evolution* 33: 402-16.

Lande, R. 1980. The genetic covariance between characters maintained by pleiotropic mutation. *Genetics* 94: 203-15.

Lewis, J. L., W. B. Lew, and J. L. Zimmerman. 1980. A nonhomogeneous anthropometric scaling method based on finite element principles. *Journal of Biomechanics* 13: 815-24.

Lofsvold, D. 1986. Quantitative genetics of morphological differentiation in *Peromyscus*. I. Tests of homogeneity of genetic covariance structure among species and subspecies. *Evolution* 40: 559-73.

Olson, E. C., and R. L. Miller. 1958. *Morphological Integration*. Chicago: The University of Chicago Press.

Richtsmeier, J. T., and J. M. Cheverud. 1986. Finite element scaling analysis of normal human craniofacial growth. *Journal of Craniofacial Genetics and Developmental Biology* 6: 289-323.

Riedl, R. 1978. *Order in Living Organisms*. New York: Wiley & Sons.

Sneath, P. H., and R. R. Sokal. 1973. *Numerical Taxonomy*. San Francisco: W. H. Freeman.

Szalay, F. S., and E. Delson. 1979. *Evolutionary History of the Primates*. New York: Academic Press.

Turelli, M. 1988a. Population genetic models for polygenic variation and evolution, pp. 601-18. In *Proceedings of the Second International Conference on Quantitative Genetics*, ed. B. S. Weir, E. J. Eisen, M. M. Goodman and G. Namkoong. Sunderland, MA: Sinauer Associates.

Turelli, M. 1988b. Phenotypic evolution, constant covariances and the maintenance of additive variance. *Evolution* 42: 1342-47.

Wagner, G. P. 1984. On the eigenvalue distribution of genetic and phenotypic dispersion matrices: Evidence for a nonrandom organization of quantitative character variation. *Journal of Mathematical Biology* 21: 77-95.

Wagner, G. P. 1988. The influence of variation and developmental constraints on the rate of multivariate phenotypic evolution. *Journal of Evolutionary Biology* 1: 45-66.

Wagner, G. P. N.d. The multivariate mutation-selection balance with constrained pleiotropic effects. Submitted.

# Key Evolutionary Innovations, Differential Diversity, and Symecomorphosis

*Karel F. Liem*

Recent studies in functional morphology, physiology, and ecology have emphasized the importance of an integrative phylogenetic approach to major biological problems (e.g., Dullemeijer 1980; Feder and Lauder 1986; Feder et al. 1988; Liem 1987; Wainwright 1987, 1988; Wake and Larson 1987; Zweers 1979). In this paper I present a tentative theoretical framework for future integrative research in organismal biology, and use the concept of *key evolutionary innovation* (KEI, Liem 1974) as an example. I propose that the origin of the KEI, its functional and morphological diversification, and possible effects on the basic way of life of the taxon possessing it, should be analyzed on the premise that organisms are participants in both the historical (reproductive, genetical) and the ecological (energy transfer) sectors. Organisms embody two great classes of interacting biological systems: one composed of units of genetic information (the historical, informational or reproductive sector), and another containing the ecological units (the economic sector or energy/matter transfer). Studies confined to just one of these sectors will, therefore, remain incomplete and may produce flawed explanations of the evolutionary process and pattern as they relate to KEIs.

The proposed theoretical framework is rooted in the hierarchical view of life as proposed by Eldredge (1985, 1986). First I review the symmetry of the historical (or genealogical) and ecological hierarchies and the concept of symecomorphosis presented elsewhere (Liem 1989). Possible implications of symecomorphosis and arguments for the role of KEIs in producing differential diversity are presented. Finally, the merits and challenges of research in the context of symecomorphosis are incorporated in the concluding remarks.

## Symmetry of Historical (Genealogical) and Ecological Hierarchies

Because organisms are participants in the historical/reproductive and the ecological/economic sectors, natural selection working in an environment of finite resources becomes an inevitability (Liem 1989). How an organism succeeds in the ecological sector has profound effects on how well it fares in its reproductive/genealogical sector. Eldredge (1985, 1986) has shown that two great general classes of biological processes resulting in two parallel hierarchies (fig. 1) act to lend cohesion to parts of any biological individual. Individuals are held together either by direct interaction of functional units or by production of offspring. The following discussion is a slightly modified version of the same topic presented elsewhere (Liem 1989).

Eldredge (1986) has argued convincingly for the recognition of hierarchies because individualized biological entities are nested, with smaller entities forming parts of larger entities, i.e., genes, chromosomes, cells, organisms, demes, species, and monophyletic taxa (fig. 1). In the genealogical hierarchy codons, genes, and chromosomes are grouped as the "germ line." Together with cells, the germ line nests into organisms; grouping of organisms produces demes, which nest into species, which in turn are grouped into monophyletic taxa. On the ecological side is the integration of functional units into organisms. Nesting of organisms produces avatars, which are local conspecific populations that interact in such a way as to produce the cohesion of local ecosystems (Damuth 1985). Avatars are integrated into local, then progressively larger-scale ecosystems and finally into the biosphere. Evolution is a dynamic interaction between energetics/ecological factors and genetics/genealogy. According to this hierarchy theory, evolution is change in genetically based information that results from processes (1) intrinsic to

GENEALOGICAL HIERARCHY    ECOLOGICAL HIERARCHY

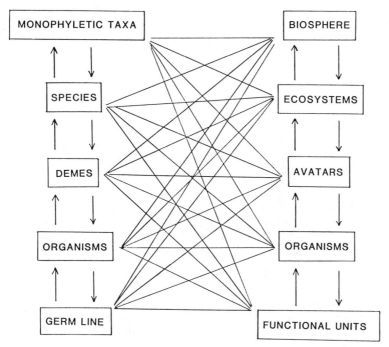

Fig. 1. The theoretical maximum of interactions between genealogical and ecological hierarchies, with upward and downward causation within each, and a complete spectrum of interhierarchical interactions. Slightly modified from Eldredge 1986.

the genealogical hierarchy of individuals within and between levels, and (2) between functional units and between individuals of the economic hierarchy. Various levels are recognized within the hierarchies, because there are processes intrinsic to each level that are not reducible to those of lower levels or subsumed by higher levels. In theory many causal interactions occur between levels within each hierarchy and a spectrum of possible causal interactions can take place between two hierarchical systems (fig. 1). Eldredge (1985, 1986) presents a conservative view of symmetrical interactions between the two hierarchies stressing that only organisms are part of both hierarchies. A major task for biologists is to analyze the nature of interactions between levels within each hierarchy and interactions between the two hierarchies.

*Karel F. Liem*

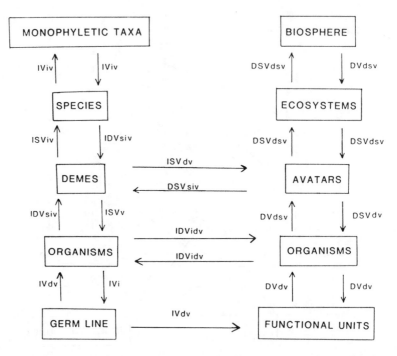

Fig. 2. Proposed influences between successive units within the genealogical and ecological hierarchies and interhierarchical influences. For explanation, see text. Abbreviations: *D*, design, broadly defined as the combined phenotypic attributes of form, behavior, and physiology and the functional properties of groups of organisms; *I*, genetic information; *S*, size; *V*, variation or diversity. Modified from Liem 1989.

Once the precise interacting processes within and between hierarchies are understood, better and more comprehensive tests can be formulated to explain the relationship between KEIs and differential diversity among lineages. An exclusive reliance on phylogenetic or historical testing is to ignore the existing cross bridge(s) between the genealogical and ecological hierarchies.

As an initial step, I propose the following analysis of the interacting processes within and between the two hierarchies (fig. 2). Adopting the convention of Dullemeijer (1959) for networks of interacting constraints, arrows flow from the influencing to the influenced component. Symbols next to the arrows indicate the influencing (in capitals, e.g., D) and the influenced (in lower case, e.g., d) properties. Thus, in the genealogical hierarchy it is the genetic information (I) of the germ line that influences the variation (v) and the design (d) of the organisms. Design is broadly defined as the combined phenotypic attributes of form, behavior, and physiology, and the functional properties of groups of organisms, species, and supraspecific categories. Ascending to the next level the design and variation of the organisms influence the size and variation of demes, which, in turn, influence the size and variation of demes, and likewise size and variation of the species influence the variation (diversity) of monophyletic taxa.

Higher levels also exert their influences on properties of the next lower levels within the genealogical hierarchy. Monophyletic taxa contain larger-scale packages of genetic information and variation with which they influence the variation and genetic information of species.

Within the ecological hierarchy we can also identify interactive influences up and down the hierarchy. At the base of the hierarchy I recognize functional units (Dullemeijer 1959, 1974; Gans 1969) which play a key role in the design and energetic attributes of the organism. The nature of the design and variation of the functional units greatly influences the design (performance in the economic arena) of the organisms, which in turn influences the size, variation and functions of the avatars. Avatars exert major influences by their design (functional attributes), internal variation, and size upon the size, function, and variation of ecosystems. Of course, ecosystems do determine the very existence of the biosphere.

As one proceeds down the ecological/energetic hierarchy one can recognize specific influences (design or function) of the next lower level. At the root of the ecological hierarchy are the functional units; the design and nature of the integration of these units determine how well the organism fares or competes in the ecological arena. The integration of functional units can be expressed as a network of interacting constraints (Dullemeijer 1974; Liem 1980; Lauder 1981). In the ecological sphere, it is the design characteristics of coupling/decoupling, developmental canalization, and spatial interactions that determine the function, variation, and behavior of the organism.

The ecological and genealogical hierarchies are mutually interdependent. The ecological/energetic side is utterly dependent upon what is supplied by the genealogical/informational side. Thus the informational side supplies the organisms or the players in the ecological (or energetic) theater. The genealogical side supplies the genetic information, variation, and design of the organisms in the ecological hierarchy. However, the genealogical side is equally dependent upon the ecological side. It is the ecological processes regulating the internal dynamics of economic systems that ultimately determine the fates of organisms in *both* hierarchies! Thus, a cross arrow is at the level of organisms from the genealogical to the ecological hierarchy. This is because the genealogical side provides organisms containing a package of information, a certain degree of variation, and a characteristic design to the ecological hierarchy where the ultimate fate of organisms is determined by ecological processes. But since evolution is defined as what happens in the genealogical hierarchy, an arrow must also be drawn in the direction of the genealogical side. It is the ecological game that ultimately determines whether organisms continue to exist in the genealogical hierarchy and, if so, it can exert a pervasive influence on the nature of the design, the variation, and the informational content of those descendants. Change and stasis in the genealogical hierarchy are regulated to a significant degree by the interacting processes of the ecological hierarchy, i.e., adaptation. However, the genealogical side is the source of genetic information and variation. Thus a balanced picture of nature emerges with a coexisting and mutually interdependent symmetrical relationship of the ecological and genealogical hierarchies. This balanced symmetry of the coexisting and mutually interdependent ecological and genealogical hierarchies in nature I call *symecomorphosis* (Liem 1989). The term has already been criticized as a grievous neologism (McKinney 1988). Unfortunately, no single existing term describes this relationship between the ecological and genealogical hierarchies. Symecomorphosis predicts that it is especially the disturbance in the ecological systems that will destroy the balanced symmetry (stasis) of the two hierarchies. Symecomorphosis also predicts that balanced symmetry can be disturbed during spontaneous modification of genealogical entities even when economic systems remain rather stable. After all, the role of the informational hierarchy is to supply functional units in a particular pattern of variation and design to the ecological arena.

I proposed (Liem 1989) that any test attempting to establish a relationship between KEIs and differential diversity of lineages must consider the principle of symecomorphosis.

## Possible Implications of Symecomorphosis

The implications of symecomorphosis cannot be fully assessed until the entire spectrum of the intra- and interhierarchical processes have been completely analyzed and understood. In the absence of such a comprehensive data base, we can make only preliminary predictions and statements concerning the possible implications of symecomorphosis on the role of KEIs in the patterns and processes of diversity.

**Evolutionary Stasis.** Symecomorphosis predicts that the reproductive success of an organism in the ecological arena and hierarchy depends on how well it interacts with the physical surroundings. *Living Fossils* (Stanley and Eldredge 1984) illustrates how the informational side of life (genealogical hierarchy) can persist almost indefinitely as long as the energetic (ecological hierarchy) systems remain undisturbed. Such living fossils as *Latimeria* occupy a specialized, geographically very restricted ecological area. Although we know very little about its physiology, *Latimeria* is undoubtedly a phylogenetic relict with functional, reproductive, and behavioral attributes, which contribute to its continued existence in a well-defined small ecological area. Its viviparous reproduction (Smith et al. 1975), the fatty swim bladder providing neutral buoyancy, the highly specialized locomotory behavior, and its massive dermal armor can be regarded as adaptations facilitating the continued existence of *Latimeria* as long as the ecological conditions do not change substantially. Evolutionary stasis is the result of a relatively stable ecological hierarchy favoring certain ways of life for which the organism is well adapted. Stabilizing selection imposed by the unchanging economic hierarchy would act against new phenotypic variants that emerge from the genealogical hierarchy. Symecomorphosis predicts that a major change in the ecological conditions leads to extinction of the living fossil which has been subjected to prolonged stabilizing selection working against the production of new phenotypic variants.

**Neutral Evolution.** Symecomorphosis also predicts that the germ line influences the design of functional units (cross bridge between germ line and functional units in fig. 2), thereby changing the functional units and consequently the organisms. Since these changes in the functional units originate in the informational factors of the germ line of the genealogical hierarchy, the result is evolution. Such changes may not affect the economic efficiency of the organisms significantly. Organisms with altered designs will, therefore, continue to operate within the ecological hierarchy. This increased diversity in form will also be reflected in an increased diversity (v in fig. 2) in the genealogical hierarchy through the interhierarchical feedback loop at the level of the organism. The result is a genetically based increase in designs even though the new variants in integrated functional units may not convey increased efficiency within the existing ecological hierarchy. As long as the variants are provided with an undisturbed ecological hierarchy, symecomorphosis will predict that they will survive, albeit not at an energetically superior level if compared with the coexisting original forms. Since evolution is in the form of changes on the genealogical side but without altering or disturbing the ecological side, this kind of symecomorphological phenomenon is called *neutral evolution*. From the perspective of an evolutionary and functional morphologist, the changes in the functional units are neutral since they do not seem to convey appreciable energetic gains within the ecological hierarchy.

**Directional Evolution and the Effect Hypothesis.** *Directional evolution* is the result of an internal propensity in a lineage producing variants of a particular form, e.g., increase in size and subsequent changes in many design parameters. Since this internal propensity is genetic, it belongs to the genealogical realm. If these newly produced variants are viable within the ecological hierarchy the branching tree of the lineage will be skewed toward the branch where the new variants are produced. The result is that a morphological trend can be recognized. Such a trend may not be strictly adaptive; it can be caused by a change in the processes on the genealogical side without altering the ecological hierarchy since the new variants function well energetically within the existing ecological arena.

But symecomorphosis predicts that nonadaptive trends are rare, since the injection of new variants with distinct design changes accompanied by alterations in energetic efficiencies would most likely alter the ecological arena. The injection of variants with economically superior designs from the genealogical to the ecological hierarchy predictably would change the interacting processes of avatars and ecosystems (fig. 2). Such an increase in specialists by higher speciation rates will subdivide ecosystems into fine-grained subunits. In the evolutionary history of related taxa with a spectrum of specialists to generalists the branching tree is again skewed and leans heavily toward the rapid specialized speciators which appear as a morphological trend. In such patterns of diversity, symecomorphosis emphasizes the role that new designs play in allowing new ways of succeeding in the ecological sphere. The skewed branching tree reflects the interactive informational influences of the genealogical hierarchy upon the design and variants of the organisms in the ecological hierarchy. The skewing toward an adaptive trend is caused to a great extent by the internal predisposition of the informational sector favoring the production of certain forms that are progressively more efficient in the transfer of matter and energy in the particular ecological hierarchy.

**Adaptive Radiation.** When the informational influences originating from the genealogical side furnish numerous variants of functional units on a common theme to an equally rich network of interactive influences in the genealogical hierarchy, the system is poised for *adaptive radiation.* Thus the variants with a wide range of designs can maximize the matter and energy transfer up and down the ecological hierarchy. The variants will in turn influence the genealogical hierarchy at the organism level. If the design of the functional units possesses a built-in versatility, the informational supply does not contain a bias toward a particular trend and a diversification of forms would be inevitable if certain processes and conditions prevail in the ecological hierarchy. Adaptive radiations of the cichlid fishes in the Great Lakes of East Africa (Fryer and Iles 1972), the Hawaiian honeycreepers (Bock 1970), and Darwin's finches on the Galapagos Islands (Grant 1986) illustrate the principle of the interplay between the informational supply, the versatility of the design of functional units and the enriching processes and features of the ecological hierarchy, especially as new energetic systems open up on islands and in lakes. It is the major goal of symecomorphosis to analyze and explain the precise nature and effects of the informational packages, the functional units and their integration, the processes between the levels up and down the two hierarchies, and the causal bridges between the two hierarchies.

**Extinction.** From the perspective of symecomorphosis the balance between the coexisting and mutually interdependent genealogical and ecological hierarchies can be disrupted in two major ways: (1) Spontaneous modification of genealogical entities may provide either the wrong players to the ecological theater or a greatly reduced variation. Thus the design and variations of the functional units are greatly influenced by the informational packages of the germ line. If variation in the functional units is reduced and the monotypic design is unable to perform adequately in the economic arena, the balance can be disrupted and cause *extinction*. (2) More commonly, the balance between genealogical and ecological hierarchies breaks down by an environmental perturbation. Abiotic factors such as a major bolide impact or temperature change will destroy the energetic organization of life, which may adversely affect the biotic world and ecosystem. Destruction of higher levels of organization within the ecological hierarchy will influence the design, size, and variation of the successive lower levels of avatars and organisms. Because of the causal cross bridges at the levels of avatars/demes and organisms/organisms, the collapse of the ecological world would inevitably lead to the extinction of the lower levels of the informational world (fig. 2). The greater the disturbance of the ecological units, the larger the genealogical units that will become extinct.

**Balanced Interhierarchical Interactions and KEIs as a Trigger of Diversity.** As discussed above, deaths of genealogical units above the organisms level are most likely caused by catastrophic events in the ecological hierarchy. Production of new forms in the genealogical hierarchy are, to a large extent, also influenced by the conditions and processes within the ecological hierarchy. The emergence of new monophyletic taxa is characterized by the acquisition of one or more phenotypic novelties, which are usually modifications of preexisting building blocks. These novelties are the synapomorphies shared only by organisms within the ancestral species and by all its descendants. When such a synapomorphy or novelty responds to a specific set of environmental conditions it may allow or trigger a radiation, and it can be classified as a KEI. The emergence of KEIs is generally a reaction to events occurring in the ecological hierarchy. Many examples of KEIs have opened up entirely new ways of life in the ecological hierarchy (e.g., Schaeffer 1948; Mayr 1959; Simpson 1959). I will discuss two examples which demonstrate the dominance of the ecological arena in the success of a taxon possessing a KEI.

Endothermal homoiothermy is the essence of being a mammal or bird. It is expressed in internal regulation of constant and high body temperatures which opened up nocturnal niches to the earliest endotherms. To be active at night was important for them to persist and operate in the ecological sphere dominated by reptiles. It can also be argued from well-documented physiological and biochemical processes that endothermal homoiothermy is necessary for effective vertebrate flight. High metabolic rates enable the central nervous system to function more rapidly, enhancing the rate of information processing and, thereby, prey capture and predator avoidance. Endothermal physiology conveys a significantly greater stamina and capacity to produce continuous muscle contractions. Endothermy also provides a greater independence of ambient temperatures. Thus endotherms are highly efficient in the economic arena, and they can operate at night when solar energy is absent and in cold climates. Based on the precise experimentally based knowledge of their physiology, behavior, and ecology, and relating these parameters with the evolutionary history of birds and mammals, the most logical conclusion is that endothermal homoiothermy has arisen twice and is a KEI which underlies the unquestionable success of birds and mammals in both their ecological and genealogical hierarchies. In this example the recognition of the KEI and its biological role is based on an extensive understanding of the informational as well as matter/energy transfer mechanisms and in relating this unequivocal data base with the evolutionary history and patterns of diversity of the two independent vertebrate lineages. Because the KEI has arisen twice independently with comparable effects on the patterns of diversity in the birds and mammals, one can conclude that the KEI is causally related with the pattern of diversity.

The second example deals with the feeding apparatus of snakes. Ophidian feeding is a truly novel way of functioning in the ecological sphere. The KEI in the ophidian feeding apparatus involves (fig. 3A, B) the decoupling of the mandibular halves at their tips, and at least eight linkages with joints between them that permit movements in various axes (Gans 1961). These multiple linkages allow a tremendous complexity of movements (fig. 3B), and since each linkage is paired, each side of the head acts independently. This KEI enables snakes to capture and swallow very large prey, thereby acquiring a new energetic position in the ecological theater (Greene 1983). In solenoglyph snakes (the true and pit vipers) a shortened mobile maxilla with a hollow fang that can be folded against the roof of the mouth when the jaws are closed is superimposed on the basic multilinkage snake design (fig. 3A, B). The fang is erected by action of the levator pterygoideus and protractor pterygoideus muscles (Dullemeijer 1959; Liem et al. 1971) which pulls the palatoquadrate arch forward and down to push the maxilla so that it is rotated (fig. 3B). The folding mechanism permits long fangs to inject venom deep into the prey. A solenoglyph snake can inject the venom and allow the prey to run off to die. The predator follows the scent trail to the dead prey. In this new mode of feeding it is no longer constrained by the size, struggle, and often effective defenses of the prey. Vipers and pit vipers, more than nonvenomous snakes, regularly feed on prey that is larger than their own body (Pough and Groves 1983). Thus solenoglyph snakes entered the economic world with a new set of highly efficient tools and diversified into major radiations by evolving numerous variations. The well-documented success of snakes in the ecological hierarchy (Greene 1983) can be related with the equally well-documented integration of functional units (Dullemeijer 1959). The economic success of snakes in the ecological hierarchy influences the informational side (fig. 2) in respect to the variational, size, and design properties, which in turn have indirect effects up and down the genealogical hierarchy. Even though this KEI has evolved only once, detailed knowledge of the processes within and between the two hierarchies represents sufficient evidence to ascribe a causal role to the KEI in shaping the ophidian and solenoglyph patterns of diversification in the ecological as well as informational hierarchies.

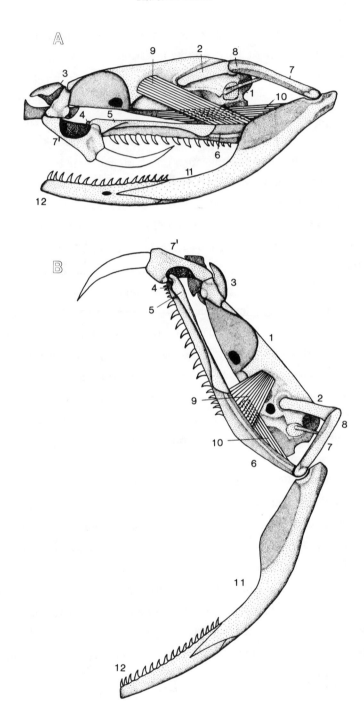

### Dominance of the Ecological over the Genealogical Hierarchy
### as a Trigger for Diversity

An evolutionary novelty may remain in a stasis for extended times when it does not convey an improvement in the matter/energy transfer. It is not until new ecological conditions and opportunities open up that the novelty expands and modifies its functional repertoire to enable new ways of life for the organism possessing it. Cichlid fishes possess such a KEI (Liem 1974). Generally river cichlids have not diversified as dramatically as the model of the KEI predicts. It was not until the ecological opportunities drastically increased in the Great Lakes of East Africa that the full potential of the KEI came to fruition in the form of dramatic radiations in form and function of the feeding apparatus (Fryer and Iles 1972; Liem 1974). This example emphasizes how dominant the ecological conditions and processes are in triggering diversity. Similarly, one can argue for the dominance of the processes of the ecological hierarchy over the genealogical side in the diversification of fringillids and drepanidids in, respectively, the Galapagos and Hawaiian Islands, in the assumed absence of a KEI in the informational hierarchy. Of course, the principle of symecomorphosis predicts that major innovations in the ecological hierarchy are as likely to trigger diversifications as innovations originating from the informational side.

The importance of ecological factors is also elucidated in the evolution of the Teleostei. A unique hydrostatic organ, the swim bladder, evolved in the Teleostei (Liem 1988), giving them a very efficient mechanism to maintain neutral buoyancy, which provides a significant economic advantage in the aquatic environment. For a swim bladder to function properly as a hy-

---

Fig. 3. Kinematics of jaw mechanism of a solenoglyph snake. A. Position of linkages, bones, and muscles with jaws closed and fangs folded. B. Position of linkages, bones, shortened muscles, and erected fang with opened jaws. Note extraordinary gape and position of the fang. Palatoquadrate is rotated forward and downward by action of levator and protractor pterygoideus muscles. Note also rotation of supratemporal and quadrate bones. *1*, braincase; *2*, supratemporal; *3*, ethmoid complex; *4*, palatine; *5*, ectopterygoid; *6*, pterygoid; *7*, quadrate; *7'*, maxilla; *8*, supratemporal-quadrate joint; *9*, levator pterygoideus; *10*, protractor pterygoideus; *11*, mandible; *12*, mandibular tips of left and right sides.

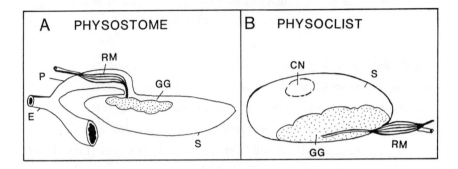

Fig. 4. Swim bladders with their respective gas-producing and gas-absorbing organs. A, physo-stomous eel swim bladder. B, typical physoclistous swim bladder. *CN*, absorbent organ; *E*, esophagus; *GG*, gas gland; *P*, pneumatic duct; *RM*, *rete mirabile*; *S*, swim bladder.

drostatic organ, its buoyant volume must be kept at ambient pressures. Physostomous fishes have open swim bladders and inflate or deflate their bladders through a pneumatic duct (fig. 4) by surfacing. Clearly the deeper its niche the greater the physical stress for a fish to regulate its buoyant volume in this way. Most marine teleosts have evolved a closed swim blad-der (physoclistous) equipped with gas-secreting structures. Elaborate gas-producing mechanisms have evolved independently within several major lineages of teleosts. Among physostomous fishes, the eels possess two gas-producing *rete mirabilia* in the walls of the pneumatic duct and a gas gland in the swim bladder (fig. 4). In the physoclistous fishes there is no trace of a pneumatic duct. They possess a gas gland and a *rete mirabile* in the wall of their closed swim bladders (fig. 4). Thus the eels and the physoclistous teleosts have converged. The emergence of closed swim bladders with built-in gas-secreting and gas-absorbing organs can be considered KEIs since they allow a new mode of life: to move up and down at neutral buoyancy without surfacing to adjust volume in their swim bladders. Eels are physostomous, but have evolved functionally identical devices to excrete and reabsorb gases by convergent evolution. Eels have independently attained the capability to adjust volumes of their swim bladders without surfacing. Equipped with these comparable KEIs, both groups of fishes engage in extensive daily vertical migrations (Marshall 1971). Under the cover of darkness many eel and physoclistous species daily visit the ocean surface. Before sunrise they move back down to their daytime living spaces. Throughout their diurnal migrations these fishes adjust the volumes of their swim bladders to maintain neutral buoyancy. These migrations coincide with those of the zooplankton, micronekton, and nekton. Thus the ability of these fishes to migrate is made possible by the KEI and provides them with the ability to enhance their energetic efficiency by taking advantage of the daily enrichment of their food pyramid levels. The groups sharing the independently evolved KEIs exhibit considerable diversity in their swim bladders and feeding structures (fig. 5). In sharp contrast the related groups with similarly specialized swim bladders but occupying other environments where vertical migrations are not necessary, do not exhibit much diversity in their swim bladder and feeding apparatus. This contrast attests to the dominant influences of the ecological upon the genealogical hierarchy.

### KEIs and Differential Diversity

Symecomorphosis predicts that KEIs trigger adaptive radiations or diversity of forms when conditions and processes in the ecological hierarchy offer many opportunities for various energy/matter transfer designs. Several historical tests have been successfully performed on various groups (see Jensen, this volume). Endothermy and patterns of diversity in birds and mammals reflect the feedback loops between the genealogical and ecological hierarchies. This prediction agrees with the adaptationist paradigm. Historical testing requires that if a KEI is of importance for the diversity of a group as a whole, then it should contribute to the diversification of each lineage within that larger group. This is rarely the case. Even within mammals and birds (and teleosts as discussed by Jensen in this volume), we find lineages lacking any kind of diversity. Symecomorphosis explains these exceptions on

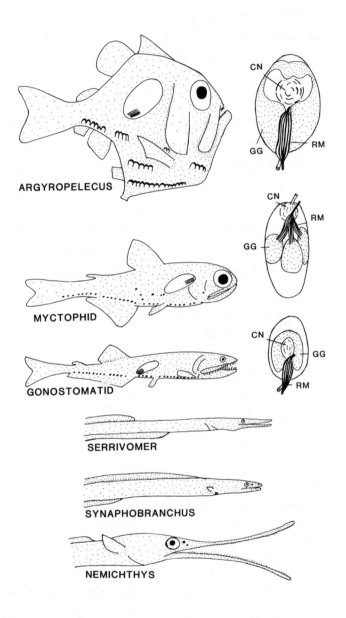

Fig. 5. Diversity of swim bladders and feeding apparatus in eels and physoclistous teleosts. Modified from Marshall 1960. *CN*, absorbent organ; *GG*, gas gland; *RM*, *rete mirabile*.

the basis of conditions and processes within the ecological hierarchy, which have been explicitly removed from consideration in historical tests. Symecomorphosis predicts that among a large diverse group with a KEI, there will invariably be lineages whose ecological hierarchies do not favor further differentiation of the KEI. This notion is adaptationist.

Significant radiations occur without a KEI as long as increased opportunities in the ecological hierarchy exist for organisms that are energetically efficient without the development of a KEI. Numerous examples among plants and animals illustrate this symecomorphological principle. This prediction is more neutral, since it states that the "good enough" (Lewontin 1978) will survive and reproduce.

Historical tests also show that there is often a great delay between the emergence of a KEI and the onset of the diversification it is assumed to have caused. If new forms are produced by the informational side in the form of new functional units and designs delivered to the ecological hierarchy, these new forms may function well under prevailing conditions and processes. But as long as the new design does not convey significant energetic advantages it will remain in a stasis until conditions and processes in the ecological hierarchy change. Only when the KEI enters a newly modified and enriched ecological hierarchy will it convey a higher efficiency in energy/matter transfer and trigger the diversification.

Symecomorphosis does not necessarily follow the adaptationist paradigm. It emphasizes that what is supplied by the informational side influences the design and variation of the functional units (fig. 2). Through developmental canalizations, decoupling of constraints, etc. (Alberch 1982; Liem 1980), the informational side profoundly affects what kind of organisms will enter the ecological theater. If the players in the ecological theater are constrained by coupled networks of interacting constraints (Liem 1980), by various developmental mechanisms (Alberch 1982), or by physiological constraints (Liem 1987) determined by the informational side, diversification will not occur regardless of how rich the economic opportunities offered on the energy transfer side. This prediction supports the notion of those with internalistic or structuralist views.

However, if the informational side supplies functional units and organisms with multiple developmental pathways and decoupled networks of interacting constraints (v in fig. 2), the organisms may diversify without major modifications or enrichment of the ecological hierarchy. The emphasis is on the intrinsic predisposition for the production of a great variation (v, V in fig. 2 lower levels of the hierarchy) in design. Diversification will then be triggered because of release of a constraint which could be a novelty as long as the new forms can be accommodated in the stable ecological arena. This prediction is in full agreement with the internalistic or structuralist notion (Wake and Larson 1987) since the diversification is not adaptive.

### Conclusion

I agree with Eldredge (1985) that "Perhaps the greatest snare and delusion of a hierarchical approach is that we come dangerously close to saying that all things are true." His statement also applies to symecomorphosis, which is based on Eldredge's view of the evolutionary process as hierarchical interactions. However, the alternative explanations of diversity as an artifact or as a stochastic phenomenon are even more ambiguous and farther removed from biology.

Organismal biology may suffer an inevitable impasse if more integrative approaches are not initiated. Symecomorphosis is the proposed integrative approach in this paper. Future challenges include: (1) Comparative and experimental studies of the nature of the integration of functional units, their developmental canalizations and underlying genetic mechanisms which lie in the realm of the historical hierarchy. (2) How functional units work, and how they perform in the economic hierarchy. (3) The assessment of the competitiveness or efficiency of the individuals within the ecological sphere.

(4) Identification and examinination of which properties of the various units within and between levels of a hierarchy interact and how these mutual interactions influence the existence and characteristics of the units. Tentatively, I propose a very coarse-grained scheme of interactions involving informational (genetic, ontogenetic) influences (fig. 2 I, i), size (fig. 2 S, s), variation (or diversity, fig. 2 S, s), and design (combined attributes of form, behavior, physiology, and functional properties of groups of organisms, fig. 2 D, d). Undoubtedly a finer-grained scheme of interactions will be discovered by experimental and broadly comparative means. Only after such a finer-grained scheme of interactive processes is documented will it be possible to test various notions explaining differential organismal diversity. (5) Systematic and evolutionary biologists will play a central role in unravelling the complex interactions within the historical hierarchy. (6) Cross bridges between the two hierarchies are still very poorly known (fig. 2). Much more research in identifying the presence and nature of the cross bridges is needed.

I propose that studies of this nature should begin with key evolutionary innovations since they potentially offer accentuated effects within and between the two hierarchies, thereby facilitating the discovery of the various interactive processes.

### Acknowledgments

I am greatly indebted to many colleagues and coworkers, all of whom have contributed significantly in the development of the various ideas and in completing this manuscript. None of them necessarily agree with every notion discussed in this paper. They include: Karsten Hartel, J. W. M. Osse, G. Zweers, Elizabeth Brainerd, Jeff Jensen, Andrew Knoll, Ernest Wu, David Smith, Christine Fox, Trish Rice, and Les Kaufman. I am especially grateful to Jeff Jensen for greatly improving this manuscript. This research was supported by NSF-BSR 8818014 and by the William Milton Fund of Harvard University.

## References

Alberch, P. 1982. The generative and regulatory roles of development in evolution, pp. 19-36. In *Environmental Adaptation and Evolution,* ed. D. Mossakowski and G. Roth. Stuttgart and New York: Gustav Fischer.

Bock, W. J. 1970. Microevolutionary sequences as a fundamental concept in macroevolutionary models. *Evolution* 24: 704-22.

Damuth, J. 1985. Selection among "species": A formulation in terms of natural functional units. *Evolution* 39: 1132-46.

Dullemeijer, P. 1959. A comparative functional anatomical study of the heads of some Viperidae. *Morphologisches Jahrbuch* 99: 881-985.

Dullemeijer, P. 1974. *Concepts and Approaches in Animal Morphology.* Assen, The Netherlands: Van Gorcum.

Dullemeijer, P. 1980. Animal ecology and morphology. *Netherlands Journal of Zoology* 30: 161-78.

Eldredge, N. 1985. *Unfinished Synthesis. Biological Hierarchies and Modern Evolutionary Thought.* New York and Oxford: Oxford University Press.

Eldredge, N. 1986. Information, economics and evolution. *Annual Review of Ecology and Systematics* 17: 351-69.

Feder, M. E., A. F. Bennett, W. Burggren, and R. B. Huey. 1988. *New Directions in Ecological Physiology.* Cambridge: Cambridge University Press.

Feder, M. E., and G. V. Lauder. 1986. *Predator-Prey Relationships: Perspectives and Approaches from the Study of the Lower Vertebrates.* Chicago: The University of Chicago Press.

Fryer, G., and T. D. Iles. 1972. *The Cichlid Fishes of the Great Lakes of Africa: Their Biology and Evolution.* Edinburgh: Boyd.

Gans, C. 1961. The feeding mechanism of snakes and its possible evolution. *American Zoologist* 1: 217-27.

Gans, C. 1969. Functional components versus mechanical units in descriptive morphology. *Journal of Morphology* 128: 365-68.

Grant, P. 1986. *Ecology and Evolution of Darwin's Finches.* Princeton: Princeton University Press.

Greene, H. W. 1983. Dietary correlates of the origin and radiation of snakes. *American Zoologist* 23: 431-41.

Lauder, G. V. 1981. Form and function: Structural analysis in evolutionary morphology. *Paleobiology* 7: 430-42.

Lewontin, R. C. 1978. Adaptation. *Scientific American* 239(3): 212-30.

Liem, K. F. 1974. Evolutionary strategies and morphological innovations: Cichlid pharyngeal jaws. *Systematic Zoology* 22(4): 425-41.

Liem, K. F. 1980. Adaptive significance of intra- and interspecific differences in the feeding repertoires of cichlid fishes. *American Zoologist* 20: 295-314.

Liem, K. F. 1987. Functional design of the air ventilation apparatus and overland excursions by teleosts. *Fieldiana (Zoology)* N.S. 37: 1-29.

Liem, K. F. 1988. Form and function of lungs: The evolution of air breathing mechanisms. Science As A Way of Knowing (SAAWOK) Symposium, American Society of Zoologists. *American Zoologist* 28: 739-59.

Liem, K. F. 1989. Functional morphology and phylogenetic testing within the framework of symecomorphosis. *Acta Morphologica Neerlando-Scandinavica* 27: 119-31.

Liem, K. F., H. Marx, and G. B. Rabb. 1971. The viperid snake Azemiops: Its comparative cephalic anatomy and phylogenetic position in relation to Viperinae and Crotalinae. *Fieldiana (Zoology)* 59: 65-126.

Marshall, N. B. 1971. *Exploration in the Life of Fishes.* Cambridge: Harvard University Press.

Mayr, E. 1959. The emergence of evolutionary novelties, pp. 349-80. In *The Evolution of Life,* ed. S. Tax. Chicago: The University of Chicago Press.

McKinney, F. K. 1988. Multidisciplinary perspectives on evolutionary innovations. *TREE* 3: 220-22.

Pough, F. H., and J. D. Groves. 1983. Specializations of the body form and food habits of snakes. *American Zoologist* 23: 443-54.

Schaeffer, B. 1948. The origin of a mammalian ordinal character. *Evolution* 2: 164-75.

Simpson, G. G. 1959. The nature and origin of supraspecific taxa. *Cold Spring Harbor Symposium Quantitative Biology* 24: 255-71.

Smith, C. L., C. S. Rand, B. Schaeffer, and J. W. Atz. 1975. *Latimeria,* the living Coelacanth, is ovoviviparous. *Science* 190: 1105-06.

Stanley, S. M., and N. Eldredge. 1984. *Living Fossils.* New York: Springer-Verlag.

Wainwright, P. C. 1987. Biomechanical limits to ecological performance: Mollusc-crushing by the Caribbean hogfish, *Lachnolaimus maximus* (Labridae). *Journal of Zoology London* 213: 283-97.

Wainwright, P. C. 1988. Morphology and ecology: Functional basis of feeding constraints in Caribbean labrid fishes. *Ecology* 69: 635-45.

Wake, D. B., and A. Larson. 1987. Multidimensional analysis of an evolving lineage. *Science* 238: 42-48.

Zweers, G. 1979. Explanation of structure by optimization and systemization. *Netherlands Journal of Zoology.* 29: 418-40.

# Plausibility and Testability: Assessing the Consequences Of Evolutionary Innovation

*Jeffrey S. Jensen*

Few biologists would doubt that an organism's characteristics might influence its success, usually thought of in terms of fitness. While there are disagreements over how these characteristics arose, both at the level of individuals (e.g., phenotypic plasticity versus genetic determinism) and lineages (e.g., adaptive change versus historical conservatism), it seems generally agreed that individual variation can profoundly influence an organism's interaction with its environment and the fate of its genetic material. Because one may in principle observe the effects of individual variation on survival or reproduction, these kinds of hypotheses may be (although usually are not) subjected to experimental tests. The extent to which attributes of lineages may influence higher level patterns is, however, a much cloudier issue. Hypotheses regarding evolutionary processes deal with entities having a complex and largely unobservable history and require different approaches for their understanding (Vermeij 1988). Historical events and their consequences are observable only indirectly, are influenced by other factors, and cannot be repeated experimentally.

Innovations characterizing clades (i.e., synapomorphies) are invoked to account for patterns of diversity, species richness, ecological breadth, or a host of other features. Terms such as "key innovation" or "adaptive breakthrough" focus on innovations which have "positive" effects, but in principle the same kinds of arguments can be applied to any outcome of evolutionary novelty. Key innovations represent only one type of relationship between evolutionary novelty and phylogenetic pattern. In examining types of arguments seeking to explain patterns of diversity, I will focus on the concept of key innovation leading to species richness simply because such arguments are so common and controversial. However, my discussion applies to historical explanations generally.

A problem that has plagued discussions of key innovations is the lack of a standard of comparison to assess diversity or species richness. Some workers (e.g., Raikow 1986) have suggested that species-rich taxa are artifacts of taxonomy and that by redrawing taxonomic boundaries we can eliminate the phenomena we are trying to explain. As an alternative to comparing groups of equivalent taxonomic rank a number of workers have suggested that comparisons of supraspecific taxa are most appropriate when made between sister groups (Lauder 1981, 1982; Cracraft 1981; Vrba 1984; Schaefer and Lauder 1986; Stiassny and Jensen 1987; Fitzpatrick 1988; Mitter et al. 1988; Raikow 1988; Vermeij 1988). Sister groups "have equivalent histories up to the time of their divergence . . . are of equal age, began with equivalent developmental programs, and differ only in those features arising (or re-emerging) after their divergence" (Stiassny and Jensen 1987). Comparison of sister groups relies not on arbitrary decisions of taxonomists to delineate groups, but on the recognition of actual units which exist in nature and have been generated by natural causes. Additional out-group comparison may be needed to determine which group exhibits the "unusual" condition, i.e., whether one group is species rich (a scenario for a key innovation in the usual sense) or its sister-group depauperate. Major differences between sister groups may reflect differences in their histories. However, identification of these differences also requires out-group comparison.

In explaining the "success" of species-rich clades, biologists have typically focused on the adaptive value of their synapomorphies. A novel feature is selected and, based on assumptions of how the derived state may be adaptively superior, a causal relationship is drawn between that novelty and the group's success. Only rarely do such adaptationist arguments explicitly relate evolutionary novelties to processes generating patterns of diversification, i.e., speciation rates, rates of morphological change, and extinction rates (Cracraft 1981, this volume). Even if such relationships are drawn, this approach has many difficulties.

A monophyletic group may have many synapomorphies, and selection from among these characters when seeking to explain patterns of diversity will reflect the researcher's bias. The radiation of cichlids provides a good example. Of the numerous features suggested to characterize the family (Liem 1973; Stiassny 1981), specializations of the pharyngeal jaw apparatus are most often cited as determinants of their success (Liem 1973). Many other synapomorphies, such as the possession of a rostral cartilaginous projection on the second epibranchial (Stiassny 1981), are not suggested to have been important in the cichlid radiation. The decision to emphasize one of many possible synapomorphies in explaining cichlid success relies on the viewpoint of the investigator. While researcher bias is unavoidable, hypotheses seeking to explain patterns of diversity should have more than their plausibility to support them. We frequently lack information about the function, if any, of characters. It seems risky to support our hypotheses merely by neglecting those systems about which we have little knowledge and appealing to the plausibility of those scenarios we choose to endorse. Plausibility is a desirable feature in such "explanations." However, it does not constitute evidence. Nor does it allow objective choice among two or more equally plausible scenarios. Decisions based on the relative plausibility of various scenarios are bound to be subjective.

Further, arguments based on the association of a unique feature with a phylogenetic outcome cannot be independently corroborated (Lauder 1981, 1982). Without independent corroboration, it is impossible to refute the possibility that differences between sister groups are due to other factors or are entirely stochastic. In the absence of independent corroboration, assessments of a causal relationship between the origin of an evolutionary novelty and patterns of diversity are either circular (the clade is speciose because it possesses this feature – the feature generated high species number as evidenced by its association with a speciose clade) or must be supported only by their plausibility.

Some workers suggest that supraspecific (or even suprapopulational) patterns are not relevant to analysis of rates of diversification, speciation, or extinction. Cracraft (this volume) views higher level patterns as "epiphenomena, or effects, of lower-level process," and argues that evolutionary innovations should be studied at or below the level of differentiating populations. Studies of the influence of evolutionary novelties at the population level will lend greater insight into how these features might influence patterns of diversification. Nevertheless, reference to actual higher-level patterns of diversity seems necessary to support hypotheses made on the basis of population-level studies. This does not support the clearly circular approach of designating a synapomorphy a key innovation simply because it characterizes a relatively speciose lineage. Rather, it asserts that hypotheses regarding causal processes must acknowledge and derive support from actual phylogenetic patterns.

### Historical Testing: An Alternative

As an alternative to adaptive arguments explaining patterns of diversity, a number of workers have advocated a procedure of historical testing (Lauder 1981, 1982; Huey 1987; Stiassny and Jensen 1987; Mitter et al. 1988; Lauder and Liem 1989). Although variously formulated, the procedure involves two main steps. In the first step, a group possessing a proposed key evolutionary novelty is compared to its sister group to assess its properties (e.g., species richness, diversity, ecological breadth). The second step investigates independent cases where the novelty has arisen to determine whether the pattern seen in step 1 is maintained. Historical testing uses convergence to seek independent corroboration of hypotheses relating evolutionary novelties to phylogenetic patterns. (See Lauder and Liem 1989 for an elaboration and formalization of this procedure.)

## Example of a Historical Test: The Labroidei/Beloniformes

One of the best known examples of a proposed key evolutionary innovation is the pharyngeal jaw specialization in cichlid fishes (Liem 1973; Stiassny and Jensen 1987; Cracraft 1981). Liem (1973) suggested that the emergence of a highly integrated pharyngeal jaw apparatus in the Cichlidae "resulted in an astonishingly dramatic episode of proliferation." In generalized percoids, the pharyngeal jaw apparatus serves primarily to transport food into the esophagus, with very little processing occurring in the pharynx (Liem 1973; Lauder 1983). There is no direct muscular connection between the neurocranium and the lower pharyngeal jaws (fig. 1A). Cichlids, in contrast, have a number of (relatively) derived features of the pharyngeal jaw apparatus. The pharyngeal jaw apparatus in cichlids differs from that in generalized percoids by the possession of united lower pharyngeal jaw elements (fig. 1C); an apophysis on the base of the neurocranium which is involved in a "true diarthrosis" with the upper pharyngeal jaws (fig. 1D); and, most importantly in Liem's opinion, a myological shift in which the lower pharyngeal jaw is suspended directly from the neurocranium by a muscular sling (fig. 1B) (Liem 1973). Primitively, the levator externus 4 originates on the lateral neurocranium and inserts on the fourth epibranchial in the dorsal gill-arch skeleton. In cichlids there is fusion of the levator externus 4 with the pars centralis of the obliquus posterior and, in some species, fusion of the levator posterior with the pars lateralis of the obliquus posterior, resulting in a continuous muscular connection between the neurocranium and the lower pharyngeal jaw (Aerts 1982; Claeys and Aerts 1984). The shift of the levator externus onto the lower pharyngeal jaw resulted in "an enormous range of possible functions that can be achieved by the total pharyngeal jaw apparatus" (Liem 1973).

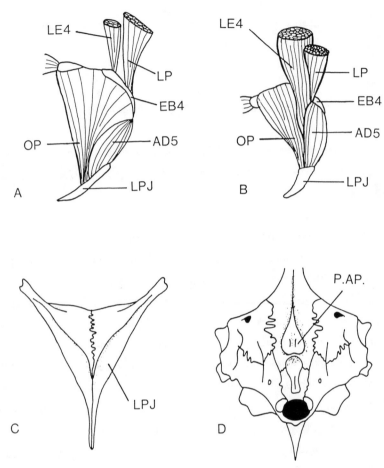

Figure 1. Perciform pharyngeal jaw apparatus. A. Isolated pharyngeal jaw musculature of *Percichthys* showing the primitive condition of the levator posterior (LP) and levator externus 4 (LE4) inserting solely on the fourth epibranchial (EB4). B. Muscle sling of *Astatotilapia* showing the derived condition of levator externus 4 insertion onto the lower pharyngeal jaw (LPJ). C. Ventral view of the lower pharyngeal jaw (LPJ) of *Astatotilapia* showing sutural union of contralateral elements. D. Ventral view of neurocranium of *Tylochromis* showing pharyngeal apophysis (P.AP). AD5, Adductor 5; OP, Obliquus posterior. Modified from Stiassny and Jensen 1987.

The development of a pharyngeal jaw apparatus capable of extensive food processing "freed the mandibular and premaxillary jaw mechanisms from their dual tasks of food collecting and preparation by eliminating the latter function" (Liem 1973). Freed from the task of food processing, the oral jaw

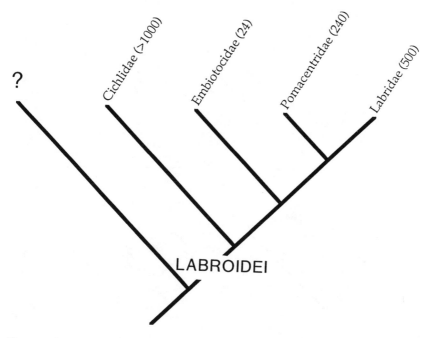

Figure 2. Intrarelationships and number of extant species of the Labroidei. Modified from Stiassny and Jensen 1987.

apparatus specialized on prey collection in ways incompatible with the former prey-processing function. To Liem the emergence of a highly integrated pharyngeal jaw apparatus was a key innovation of such adaptive potential that it resulted in the extreme diversification and species richness of the Cichlidae. The above pharyngeal specializations have since been shown to characterize a much more extensive group – the Labroidei (fig. 2) (Stiassny 1980; Liem and Greenwood 1981; Kaufman and Liem 1982; Stiassny and Jensen 1987). The Labridae, Embiotocidae, Pomacentridae, and Cichlidae share united lower pharyngeal jaw elements (also found elsewhere in the Percomorpha, usually only in durophagous species), a basicranial apophysis with a true diarthrosis with the upper pharyngeal jaws, and a muscle sling supporting the lower pharyngeal jaw. The muscle sling is polymorphic within the Pomacentridae but for the purposes of this discussion is assumed to be a synapo-

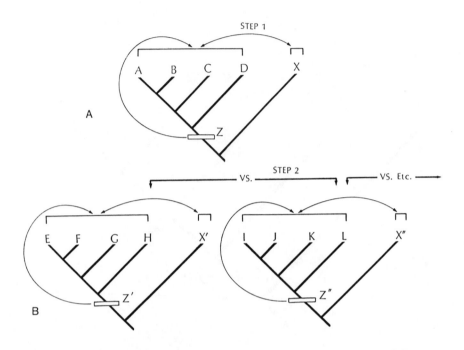

Figure 3. Two-step process of historical testing. A. Step 1: Comparison of lineage exhibiting an evolutionary novelty (Z) with its sister group. B. Step 2: Comparison of sister groups in independent lineages in which the "same" novelty (Z', Z") has arisen. From Stiassny and Jensen 1987. Reprinted with permission of Museum of Comparative Zoology, Harvard University.

morphy for the Labroidei as a whole. (See Stiassny and Jensen 1987 for more details.)

A historical test of the importance of labroid pharyngeal jaw specialization in labroid species richness begins with an assessment of how species rich the group is (fig. 3A). The Labroidei contain approximately 1800 species, approximately 8% of all teleosts. Probably no single perciform group or collection of perciform groups likely to be sister to the Labroidei will match it in species number. Thus, the association of these pharyngeal specializations with species richness applies to labroids generally, not just cichlids.

The second step of historical testing, examination of repeated occurrences of the evolutionary novelty for common patterns of diversity (fig. 3B),

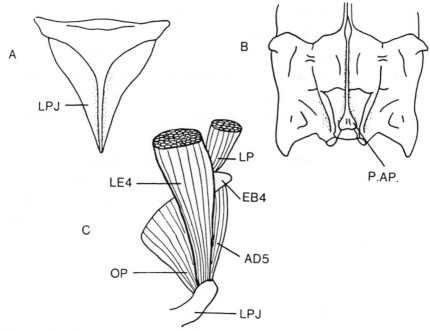

Figure 4. The beloniform pharyngeal jaw apparatus. A. Ventral view of the lower pharyngeal jaw (LPJ) of *Exocoetus* showing fusion of contralateral elements. B. Ventral view of neurocranium of *Exocoetus* showing pharyngeal apophysis (P.AP.). C. Muscle sling of *Exocoetus* showing insertion of levator externus 4 (LE4) onto the lower pharyngeal jaw (LPJ). AD5, Adductor 5; EB4, Epibranchial 4; LP, Levator posterior; OP, Obliquus posterior. Modified from Stiassny and Jensen 1987.

poses some difficulty. This step requires the evolutionary novelty to have arisen independently in other lineages. In the case of labroid pharyngeal specializations, we are fortunate in having another group, the Beloniformes, within which many of the pharyngeal specializations described for the Labroidei are mirrored to a remarkable degree (Stiassny and Jensen 1987; Rosen and Parenti 1981). The Exocoetoidei are characterized by a fusion of the lower pharyngeal jaw elements (fig. 4A), as opposed to the close apposition seen in adrianichthyoids (Rosen and Parenti 1981). More remarkably, the Exocoetoidea (Hemirhamphidae plus Exocoetidae) possess a direct articulation of the upper pharyngeal jaws with a neurocranial apophysis (formed from the basioccipital rather than the parasphenoid as in labroids; fig. 4B) and a muscle sling consisting of the levator externus 4 extending to and supporting the lower pharyngeal jaw (fig. 4C).

The independent, yet remarkably similar, development of pharyngeal specializations in these two lineages allows an independent assessment of their effects on species richness. If Liem (1973) is correct, the Exocoetoidea should be more speciose than its sister group, the Scomberesocoidea, because the Exocoetoidea possesses a muscle sling and basicranial articulation. The Exocoetoidea is indeed more speciose than the Scomberesocoidea (135 + species versus 36; fig. 5). Thus, Liem's hypothesis is supported by the repeated (albeit only once) association of this structural complex with increased species number. This is an improvement over unsupported arguments about adaptive value.

## Problems and Requirements of Historical Testing

Historical testing has been widely discussed recently due to both increased interest in evolutionary innovations (e.g., this volume) and in more explicit appeal to phylogeny in interpreting evolutionary patterns (e.g., Liem and Wake 1985; Wake and Larson 1987; Huey 1987; Schaefer and Lauder 1986; Mitter et al. 1988). However, historical testing poses a number of difficulties. For many taxa the sister groups will be either unclear or completely unknown, rendering comparisons ambiguous (Mitter et al. 1988; Cracraft, this volume). This is an empirical rather than methodological problem, which underscores the importance of phylogenetic information for interpreting patterns of diversity (Cracraft 1985). Extinction will also hamper application of historical tests. In an assessment of species richness, for example, extant taxa represent only a sample of the lineages that actually existed in each group (Cracraft, this volume).

In addition, historical testing makes more assumptions and requires more systematic information than simply knowledge of sister-group relation-

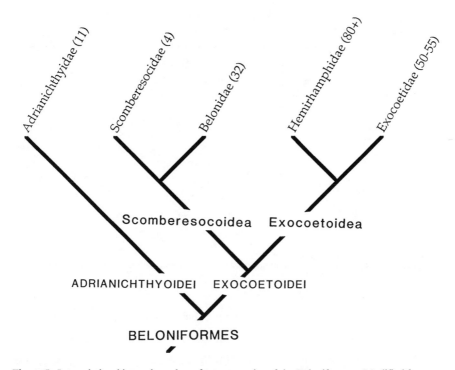

Figure 5. Intrarelationships and number of extant species of the Beloniformes. Modified from Collette et al. 1984.

ships. Although sister groups are equivalent at their origin, they later differ in ways other than the possession of a putative key innovation. Other independently derived novelties accumulate in the sister clades and subgroups within them which might unpredictably influence phylogenetic patterns (Huey 1987; Stiassny and Jensen 1987; Raikow 1988). The effects of a key innovation might be swamped out by other differences arising between sister clades. Likewise, similar evolutionary novelties arising in distantly related lineages occur in entirely different developmental, and possibly environmental, contexts and will not necessarily have the same effect. Such differences may render historical tests ambiguous (Stiassny and Jensen 1987; Raikow 1988).

Finally, information is required on phylogenetic relationships within the groups proposed to possess a key innovation. For example, the diversifi-

Ginglymodi (7)　　　　　HALECOSTOMI (>22,000)

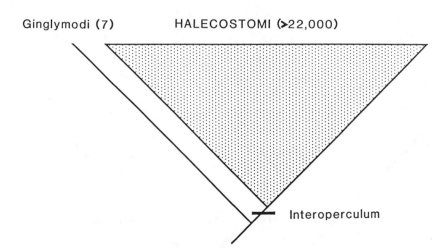

Interoperculum

Figure 6. Comparison of species number of the Halecostomi, which possesses an interoperculum, and the Ginglymodi.

cation of halecostome fishes (*Amia* plus the Teleostei extant) has been attributed to the development of an interoperculum. This innovation resulted in a new mode of mouth opening, the levator operculi coupling, that was "central to the exploitation of new and diverse trophic resources" (Liem 1980). The radiation of the halecostomes seems to be a perfect example of the role of an evolutionary novelty in determining patterns of diversity. The extant sister group of the Halecostomes, the Ginglymodi, is not speciose and probably never was (Wiley 1976). Halecostomes, on the other hand, are far more species rich (22,000 versus seven extant species; fig. 6). Thus, the halecostomes seem to be a group characterized by a key evolutionary novelty.

Examining relationships within the halecostomes weakens this argument greatly (fig. 7). If an innovation is important for the diversification of a group as a whole, it should contribute to diversification of each of the lineages within that larger group. The phylogeny of the group should be more of a bush than a ladder, with sister taxa within the group approximately equal in species richness. This is clearly not the case for the Halecostomi. While all halecostome groups are more speciose than the Ginglymodi (with the ex-

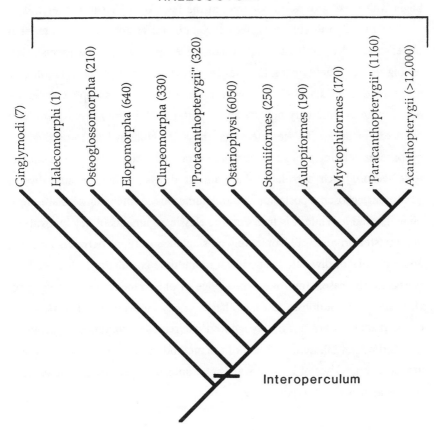

Figure 7. Comparison of the Halecostomi and the Ginglymodi showing halecostome intrarelationships. Modified from Lauder and Liem 1983.

ception of the Halecomorphi – with only *Amia calva* extant), their patterns of diversity do not suggest a group whose success can be attributed to a feature characterizing it. The statement that teleosts represent "the greatest radiation known among vertebrates" (Liem 1980) obscures the fact that this radiation has been primarily confined to the Ostariophysi and the Acanthopterygii, which together comprise approximately 85% of all teleost species (fig. 7). The search for characters contributing to teleost species richness should begin with these groups, rather than at the level of the Acanthopterygii.

Cracraft (this volume) rejects the comparative argument for a number of reasons. He points out that comparative approaches such as historical testing do not identify causation per se, but rather multiple correlation (see also Lauder and Liem 1989). However, the existence of multiple correlation between an evolutionary novelty and a pattern of diversity at the very least bears examination. To the extent that these correlations lead to predictions, hypotheses relating them are testable. Cracraft further criticizes historical tests because "such 'correlations' are almost always spurious because diverse 'groups' are chosen a posteriori . . . and numerous counterexamples are inevitably ignored." While it is true that the classic arguments of key innovations are a posteriori and, still worse, rely on single associations between a novelty and a pattern of diversity (as opposed to multiple correlation), this need not always be the case. Improved knowledge of phylogenetic relationships provides new information to test hypotheses of key innovations. The view that counterexamples are "inevitably ignored" addresses a problem of misapplication and not an inherent problem of the method. Certainly, if historical tests are used in analyses of key innovations, counterexamples should be acknowledged as contradictory evidence.

## Unexpected Patterns

Because so many factors may confound historical testing, even generally supported hypotheses have counterexamples. The Embiotocidae is a clearly monophyletic labroid subgroup that is very depauperate in species number relative to its sister group (24 species versus 800+ for the Pomacentridae/Labridae; fig. 2). Nevertheless, the Embiotocidae possess all of the pharyngeal specializations postulated to have made the labroids as a whole so speciose. If the hypothesis that pharyngeal specialization leads to species richness is to be maintained, low species number in the Embiotocidae must be explained. By analogy with traditional approaches to explaining diversity, one could examine the embiotocids for unusual characters that might explain their low species number. For example, embiotocids differ from other labroids in being viviparous. While one could construct an argument relating viviparity to low species number, this is again simply an a posteriori argument that lacks independent support. Little is gained from comparative historical tests if every counterresult is explained ad hoc (Raikow 1986). There are many other features characterizing embiotocids (e.g., pharyngobranchial 4 toothplate form, adductor mandibulae 1 configuration, shelf on the interoperculum) which, depending on the perspective of the investigator, could be considered responsible for low species number.

## Multiple Historical Tests

An alternative to such ad hoc arguments applies the historical testing approach to counterexamples as well. If viviparity inhibits species richness in the Embiotocidae, it should have similar effects in other groups. Viviparity has arisen within the Teleostei a number of times (Wourms 1981), allowing a comparison of its correlates in a variety of phylogenetic and environmental contexts. For some viviparous lineages systematic information is either lacking or is so ambiguous that the origin of viviparity cannot be assigned to a particular node (e.g., zoarcids). In other groups, the sister group of the viviparous clade is not clear. Parenti (1981), for example, placed the viviparous Poeciliidae with its approximately 140 species in a trichotomy with monotypic *Fluviphylax* (Roberts 1970) and a much larger assemblage closer in size to the poeciliids. Because the relationships are uncertain, assessing consequences of viviparity using this group seems unwarranted. However, other groups provide less ambiguous opportunities for comparison. Goodeids are a viviparous monophyletic assemblage containing 34 species, whereas their oviparous sister group, *Empetrichthys* plus *Crenichthys,* contains only six species (Parenti 1981; Nelson 1984). The viviparous monophyletic lineage *Anableps* plus *Jenynsia* has 6 species, whereas its oviparous sister lineage, *Oxyzygonectes,* is monotypic (Parenti 1981; Nelson 1984). These cases indicate that viviparity actually enhances the number of species in a lineage, although each of these groups is quite small. In other groups, this situation is reversed. For example, Cohen and Nielson (1978) divide the Ophidiiformes into two subfamilies, the oviparous Ophidioidei with 200 species, and the viviparous Bythitoidei with 100 species. Although the authors emphasize that this is a classification rather than a phylogeny, they suggest that viviparity is derived only once in the group and that the Bythitoidei are monophyletic. Likewise, the viviparous family Parabrotulidae (containing only 2 species) is depauperate relative to any of the groups to which it has been allied.

The viviparity example illustrates two important points. First, it is possible to find repeated instances of an evolutionary innovation to assess its effects independently. Groups with reasonably well resolved phylogenies are particularly amenable to this. Second, even with multiple historical tests the results may be ambiguous.

In addition to the problems inherent in the single test situation, multiple historical testing carries the added problem of the potential of infinite regress. Multiple testing may itself yield ambiguous results that either beg ad hoc explanation or still further historical testing. Nevertheless, multiple testing may partly address Raikow's (1986, 1988) lament that unequal degrees of success in independent lineages "sharing" an innovation will simply be dismissed as a result of their dissimilarity.

### Conclusion: Historical Testing as a General Comparative Method

Historical testing is becoming more popular as nonsystematists begin to incorporate more phylogenetic information into their studies. Historical testing is not limited to studies of key innovations, but applies equally well in studies of adaptation (Coddington 1988), physiology (Huey 1987), ecology, behavior (McLennan et al. 1988), and development (Alberch 1980, 1982). Although the evidence is correlative, this approach can greatly strengthen hypotheses seeking to explain current patterns of diversity.

The original dissatisfaction with the concept of key innovations revolved around the untested or untestable nature of the arguments usually proposed. Explanations based on historical testing, on the other hand, suffer from two major problems: (1) only rarely can this type of testing be applied, either because relevant phylogenetic information (regarding relationships both within and beyond the clade in question) is lacking or because some innovations may be truly unique, and (2) even where relevant comparisons can be made, the results may be ambiguous. Still, historical testing provides a nonarbitrary, data-dependent means of assessing the consequences of evolutionary innovation that does not rely on subjective adaptive arguments.

## Acknowledgments

I thank Beth Brainerd, Karel Liem, Ernest Wu, and three reviewers for helpful comments on early drafts of this paper. My interest in this subject developed from discussions with Melanie L. J. Stiassny, and I thank her for her inspiration and good humor. The cladograms were drawn with MacClade, which was generously provided by David and Wayne Maddison.

## References

Aerts, P. 1982. Development of the musculus levator externus IV and the musculus obliquus posterior in *Haplochromis elegans* Trewavas, 1933 (Teleostei: Cichlidae): A discussion of the shift hypothesis. *Journal of Morphology* 173: 225-35.

Alberch, P. 1980. Ontogenesis and morphological diversification. *American Zoologist* 20: 653-67.

Alberch, P. 1982. Developmental constraints in evolutionary processes, pp. 313-32. In *Evolution and Development*, ed. J. T. Bonner. Berlin: Springer-Verlag.

Claeys, H., and P. Aerts. 1984. Notes on the compound lower pharyngeal jaw operators in *Astatotilapia elegans* (Trewavas), 1933 (Teleostei: Cichlidae). *Netherlands Journal of Zoology* 34: 210-14.

Coddington, J. 1988. Cladistic tests of adaptational hypotheses. *Cladistics* 4: 3-22.

Cohen, D. M., and J. G. Nielson. 1978. Guide to the identification of genera of the fish order Ophidiiformes with a tentative classification of the order. *NOAA Technical Report NMFS Circular* 417.

Collette, B. B., G. E. Gowen, N. V. Parin, and S. Mito. 1984. Beloniformes: Development and relationships. *In Ontogeny and Systematics of Fishes*, ed. H. G. Moser. *Special Publications of the American Society of Ichthyologists and Herpetologists* No. 1. Lawrence, KS: Allen Press.

Cracraft, J. 1981. The use of functional and adaptive criteria in phylogenetic systematics. *American Zoologist* 21: 21-36.

Cracraft, J. 1985. Biological diversification and its causes. *Annals of the Missouri Botanical Garden* 72: 794-822.

Fitzpatrick, J. W. 1988. Why so many passerine birds? A response to Raikow. *Systematic Zoology* 37: 71-76.

Huey, R. B. 1987. Phylogeny, history, and the comparative method. In *New Directions in Physiological Ecology*, ed. M. E. Feder, A. F. Bennett, W. W. Burggren, and R. B. Huey. New York: Cambridge University Press.

Kaufman, L., and K. F. Liem. 1982. Fishes of the suborder Labroidei (Pisces: Perciformes): Phylogeny, ecology, and evolutionary significance. *Breviora* (Museum of Comparative Zoology, Harvard University) 472: 1-19.

Lauder, G. V. 1981. Form and function: Structural analysis in evolutionary morphology. *Paleobiology* 7: 430-42.

Lauder, G. V. 1982. Historical biology and the problem of design. *Journal of Theoretical Biology* 97: 57-67.

Lauder, G. V. 1983. Functional design and evolution of the pharyngeal jaw apparatus in euteleostean fishes. *Zoological Journal of the Linnean Society of London* 77: 1-38.

Lauder, G. V., and K. F. Liem. 1983. The evolution and interrelationships of the actinopterygian fishes. *Bulletin of the Museum of Comparative Zoology* 150: 95-197.

Lauder, G. V., and K. F. Liem. 1989. The role of historical factors in the evolution of complex organismal functions. Dahlem Conference, West Berlin, August 1988. Chichester, England: John Wiley.

Liem, K. F. 1973. Evolutionary strategies and morphological innovations: Cichlid pharyngeal jaws. *Systematic Zoology* 22: 425-41.

Liem, K. F. 1980. Acquisition of energy by teleosts: Adaptive mechanisms and evolutionary patterns. In *Environmental Physiology of Fishes,* ed. M. A. Ali. New York: Plenum.

Liem, K. F., and P. H. Greenwood. 1981. A functional approach to the phylogeny of the pharyngognath teleosts. *American Zoologist* 15: 83-101.

Liem, K. F., and D. B. Wake. 1985. Morphology: Current approaches and concepts. In *Functional Vertebrate Morphology,* ed. M. Hildebrand, D. M. Bramble, K. F. Liem, and D. B. Wake. Cambridge: Harvard University Press.

McClennan, D. A., D. R. Brooks, and J. D. McPhail. 1988. The benefits of communication between comparative ethology and phylogenetic systematics: A case study using gasterosteid fishes. *Canadian Journal of Zoology* 66: 2177-90.

Mitter, C., C. B. Farrel, and B. Wiegmann. 1988. The phylogenetic study of adaptive zones: Has phytophagy promoted insect diversification? *American Naturalist* 132: 107-28.

Nelson, J. 1984. *Fishes of the World.* Toronto: John Wiley.

Parenti, L. 1981. A phylogenetic and biogeographic analysis of cyprinodontiform fishes (Teleostei, Atherinomorpha). *Bulletin of the American Museum of Natural History* 168: 335-557.

Raikow, R. J. 1986. Why are there so many kinds of passerine birds? *Systematic Zoology* 35: 255-59.

Raikow, R. J. 1988. The analysis of evolutionary success. *Systematic Zoology* 37: 76-79.

Roberts, T. R. 1970. Description, osteology and relationships of the Amazonian cyprinodont fish *Fluviphylax pygmaeus* (Myers and Carvalho). *Breviora* (Museum of Comparative Zoology, Harvard University) 347: 1-28.

Rosen, D. E., and L. R. Parenti. 1981. Relationships of *Oryzias,* and the groups of atherinomorph fishes. *American Museum Novitates* 2719: 1-25.

Schaefer, S. A., and G. V. Lauder. 1986. Historical transformation of functional design: Evolutionary morphology of the feeding mechanism in loricarioid catfishes. *Systematic Zoology* 35: 489-508.

Stiassny, M. L. J. 1980. *The Anatomy and Relationships of Two Genera of African Cichlid Fishes.* Ph.D. diss., University of London.

Stiassny, M. L. J. 1981. The phyletic status of the family Cichlidae (Pisces, Perciformes): A comparative anatomical investigation. *Netherlands Journal of Zoology* 31: 275-314.

Stiassny, M. L. J., and J. S. Jensen. 1987. Labroid intrarelationships revisited: Morphological complexity, key innovations, and the study of comparative diversity. *Bulletin of the Museum of Comparative Zoology* 151: 269-319.

Vermeij, G. 1988. The evolutionary success of passerines: A question of semantics? *Systematic Zoology* 37: 70-72.

Wake, D. B., and A. Larson. 1987. Multidimensional analysis of an evolving lineage. *Science* 238: 42-48.

Wiley, E. O. 1976. Phylogeny and biogeography of fossil and Recent gars (Actinopterygii: Lepisosteidae). *University of Kansas Museum of Natural History Miscellaneous Publications* 64: 1-111.

Wourms, J. P. 1981. Viviparity: The maternal-fetal relationship in fishes. *American Zoologist* 21: 473-515.

# Studying Physiological Evolution: Paradigms and Pitfalls

*Warren W. Burggren and William E. Bemis*

**Evolutionary Physiology: Scope, Audience, Definitions, and Current Status**

The last thirty years have produced an active and growing field of evolutionary morphology which is successfully synthesizing functional anatomy and systematics (Liem and Wake 1985; Northcutt 1986; Herring 1988). While outgrowths of biochemical physiology have had a strong evolutionary impact (e.g., evolution of globin structure in Goodman et al. 1987), development of what may be termed "evolutionary physiology" – the broad study of the evolutionary origins and patterns of change in physiological systems at the tissue, organ, and organismal level (Prosser 1986) – has been slow.   D. Dwight Davis (1958) described the proper goal of comparative anatomy, stating that "The central problem of historical biology is to explain, not to describe, biological conditions that exist today." We strongly feel that the "proper goal of evolutionary physiology" is identical, and we must emphasize not just how physiological processes evolved but how organisms evolved. Who is concerned with this goal, and how is it to be achieved?

Physiologists using micro- and macroevolutionary approaches have both contributed to this goal. Ecological physiologists in particular are beginning, with great success, to adapt quantitative genetic approaches to the study of evolutionary change in physiological characters (see Feder et al. 1987 for numerous examples). These approaches, though yielding important information, reveal for the most part only *small changes within existing biological systems*. For example, Huey (1987) and Arnold (1987) have recently addressed the consequences of the quantitative analysis of physiological traits such as body-temperature regulation and locomotor performance, but their studies do not consider the origin of endothermy or locomotion. To address broader problems of the *origin* of new systems, a macroevolutionary perspective may be fruitful. By focusing on the presence or absence of systems, we draw attention to phylogenetic origins of characters rather than subsequent evolutionary changes they may show. We do not suggest that the qualitatively different features we think of as macroevolutionary change are non-Darwinian (Charlesworth et al. 1982), but rather that there is an important difference in approach and perspective in studying these features. *Analysis of evolutionary innovation* in physiology at present seems to require emphasis and understanding of qualitative characters, not extrapolation from microevolutionary models. By choosing to discuss qualitative characters, we ignore the vast literature on interspecific allometry of physiological processes. Consequently, the focus of this chapter will be at the transspecific or macroevolutionary level.

Evolutionary physiology centers on "transformational" analyses, which attempt to reconstruct evolutionary pathways using phylogenetic data, rather than on "equilibrium" analyses, which examine correlations between traits in extant animals (Lauder 1982). We cannot improve upon Huey's (1987) recent treatment of the differences between these perspectives, so we suggest that readers familiarize themselves with this important analysis. Essentially, our message is not for those physiologists (especially ecological physiologists) who practice any component of contemporary evolutionary biology. Rather, we aim at the far larger group of "mechanism-oriented comparative physiologists" (Burggren 1987a) who wish to interpret their data in a more rigorous evolutionary context. We also direct our message to systematists who believe that physiology plays only a minor role in evolutionary biology.

Having defined our scope and intended audience, we emphasize that the slow development of evolutionary physiology has not resulted from a lack of interest by comparative physiologists in evolutionary processes. Understanding the evolution of physiological characters, in addition to categorizing them, has been an implicit (though rarely explicit) goal of much twentieth-century comparative physiology (see Burggren 1989 for review). Unfortunately, comparative physiology traditionally has been, and continues to be, outside a framework of contemporary evolutionary biology, often embracing theories, positions, or approaches that contemporary morphologists, evolutionary biologists, and geneticists have abandoned (Mayr and Provine 1980; Bennett 1987; Burggren 1989).

### The Challenge of Evolutionary Physiology: Pitfalls

Why has the study of physiological evolution lagged behind that of morphological evolution? We identify here several important problems – some perceived, some real – and will discuss each in turn.

**The Nature of Structure-Function Relationships.** Functional morphologists generally regard structure and function as very tightly linked (see Lauder 1982). Many physiologists find this attitude perplexing because their experiences suggest that physiological processes are very loosely linked to the morphological structures from which they emanate. Minute differences in anatomy can and do yield great differences in physiological processes. As examples of this loose linkage, the patterns of blood pressure generated during heart contraction in reptiles can differ greatly with only small changes in cardiac anatomy between closely related taxa (Burggren 1987b, and below). Minor changes in epithelial structure can convert a region of the gut from a digestive function to a respiratory function in air-breathing fishes (McMahon and Burggren 1987). Subtle changes in the structure of the walking legs of certain intertidal crabs can change these organs from locomotory to respiratory structures (fig. 1). On the other hand, homologous organs may show great morphological differences without qualitative changes in physiological

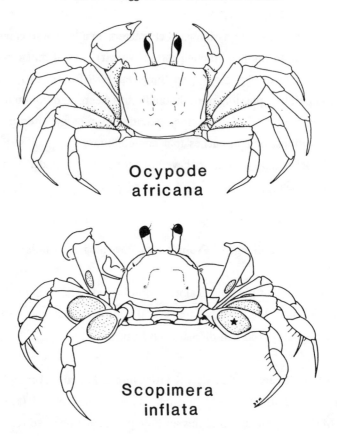

Figure 1. Examples of the loose linkage of structure and function. Two species of ocypid crabs, in which small morphological changes in structure result in major functional differences. In *Ocypode*, the legs are used strictly for locomotion. In *Scopimera*, a relatively minor morphological change involving thinning of the chitinous cuticle on a small region of the posterior surface of the walking legs (star) allows the legs to serve as respiratory organs. Modified from Maitland 1986 and Hartnoll 1988.

function. The parabronchial lungs of birds, ventilated in a unidirectional fashion using a series of air sacs, and the alveolar lungs of mammals, ventilated in a tidal fashion using a diaphragm, differ considerably in structure and mechanism. Yet, both ultimately produce the same effect – full oxygen saturation of the arterial blood (fig. 2). Other examples, such as patterns of contraction of muscles used in teleost feeding (Liem 1978) or tetrapod locomotion on land or in the air (Goslow et al. 1989), can appear conservative

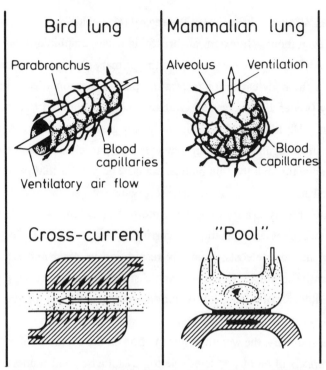

Figure 2. A comparison of the parabronchial lungs of birds and the alveolar lungs of mammals. The top panels indicate schematically the patterns of gas flow through the functional unit of the lung, while the two lower panels show the patterns of gas flow (stippled regions) and blood flow (diagonal lines). Clearly, there are great differences in lung anatomy and patterns of blood and gas flow. Yet, both systems achieve an identical function – oxygenation of pulmonary venous blood. Modified from Scheid 1982.

despite anatomical rearrangements of the muscles.

As a consequence of the generally loose association between physiological processes and readily observed structures, knowledge of how morphological characters evolved in a specific lineage may be of limited use in understanding the physiological transitions in that lineage's physiological traits. It follows that *there is no reason to suppose that the course of physiological evolution of a structure should closely parallel its anatomical evolution, and vice versa.* Consider the putative course of evolution, both anatomical and physiological, of fish swim bladders, the subject of a classical debate as to whether buoyancy or respiration is the primitive physiological character (Romer 1972;

Randall et al. 1980; Little 1983). We regard this debate as unresolvable, and possibly even moot. In extant air-breathing fishes, amphibians, and aquatic reptiles, any organ that has arisen for the primary purpose of incorporating gas within the body cavity for respiratory purposes automatically influences buoyancy (Feder 1984; Gee and Graham 1978; Lannoo and Backman 1984; Burggren 1988; Milsom 1975). Similarly, any gas brought into a thin-walled gas bladder to regulate buoyancy may very well participate to some extent in gas exchange through the thin bladder walls. The point is that swim bladders could well have arisen as dual-purpose organs (this seems almost unavoidable), with various taxa showing subtle morphological changes that ultimately led to subsequent specializations occurring in derived lineages (e.g., strictly buoyancy in most teleosts; primarily respiration in *Amia* and *Lepisosteus*). With such loose linkage between bladder structure and function, the history of the bladder's physiological evolution would be a poor predictor of its morphological history, and vice versa.

In summary, the wealth of information on the evolution of morphological features is often of restricted use to physiologists, and renders the study of evolutionary physiology much more difficult. The equivocal relevance of morphological information now leads us to discuss a related pitfall in the study of evolutionary physiology – recognition of "physiological homology."

**Recognizing Physiological Homology.** A direct consequence of the often uncertain link between structure and function is that concepts of physiological homology are often problematic or uncertain.[1] Without dealing in depth with this pitfall, we simply note several problems associated with recognition of physiological homology. These problems, perhaps a major reason why evolutionary physiology has lagged, suggest that a full, modern theoretical treatment of homology from a physiological perspective is desirable. Indeed, Coddington (1988) suggests that "discussions of homology of function . . . could become meaningful and commonplace."

Evidence that physiological homology may be a problem comes from considering comparative endocrinology, one area of evolutionary physiology that is flourishing (Vigna 1985). Perhaps the chief characteristic of comparative endocrinology is that evolutionary questions have for many years been studied by direct comparisons of peptide or nucleic-acid sequences. Because of the focus on sequences, homology recognition is simplified. As a result of being able to state homologies clearly, comparative endocrinologists have been able to contribute to evolutionary questions (e.g., Wallis 1975).

One reason why physiologists have problems with homology is that there are no easy means to assess the homology of the quantitative features of greatest familiarity and interest to physiologists except by reference to the morphological substrates of these functions. Although quantitatively varying features can be compared in a variety of taxa using various numerical methods (Felsenstein 1985; Huey 1987), this still does not solve the issue of homology recognition.

Functional biologists readily acknowledge that homologous structures may be the substrate for two or more nonhomologous functions. But can a function be homologous even if it is carried by two nonhomologous structures? Consider a nonphysiological example: vertebrate gastrulation. Patterns of gastrulation in vertebrates are extremely diverse, involving such different tissue mechanics and cell locomotor capabilities that from a purely structural level of inquiry it can be difficult to recognize the similarities that occur if the comparison spans a large phylogenetic gap. Even within a closely monophyletic group such as anurans, strikingly different morphological arrangements occur during gastrulation of *Xenopus* versus *Rana* (Keller 1981). Yet few doubt that gastrulation is a homologous process in all vertebrates. And, importantly, we do this without requiring any specific set of morphological features as the "gastrulation substrate."

Evolutionary physiologists commonly encounter certain problems of homology recognition which do not normally plague evolutionary morphologists. One of these is paralogy, the existence of genes duplicated during phylogenetic history (Patterson 1987). For example, during development, successive embryonic, fetal, and adult hemoglobins are produced by red blood cells (Huehns et al. 1964). Different, but very similar, globin genes (perhaps originating as duplicated sequences) are activated sequentially over the period of development. This pattern of "gene switching" is well known and common for many gene products used during development, such as sea urchin histones (Gilbert 1988).

An electrophysiological example demonstrates the problem of paralogy and differential tissue expression. In this case, the protein is a tetrodotoxin-sensitive, voltage-dependent sodium channel expressed in the muscle cells of chaetognaths, the arrow worms (Schwartz and Stühmer 1984). Voltage-dependent sodium channels are used in both neurons and skeletal muscle of vertebrates; in contrast, while invertebrate axons use sodium channels, invertebrate skeletal muscles typically use an inward calcium flux to generate the action potential signalling contraction. At this time we do not know whether the sodium channels expressed in chaetognath muscle and nerve cells are homologous (i.e., products of the same gene). If they are, then the evolutionarily derived character of chaetognaths is to express the sodium channel in a new tissue type – muscle cells. If they are paralogous, then we must survey other taxa to determine whether duplication of the gene for sodium channels is apomorphic for chaetognaths or a synapomorphy of chaetognaths and some larger group of organisms.

A second hypothetical example may clarify the problem. Fishes regulate salt balance in part with epithelia specialized for excretion of chloride ions. Typically, chloride ion excretion occurs at one of two sites: the digitiform gland in elasmobranchs (and lungfishes – see Lagios and McCosker 1977) or the gill epithelium of teleosts (Payan et al. 1984). We do not know whether these chloride pumps are homologous gene products, but if we assume that they are, then physiological homologues are differentially expressed by nonhomologous tissues (digitiform gland and gill epithelium). It will be a major task for evolutionary physiologists to establish whether gene products are homologous or paralogous.

Clearly, the accurate recognition of homologous processes is at the center of rigorous studies in evolutionary physiology.

**The Lack of Fossil Record.** Regrettably, the physiological fossil record is almost nonexistent, a situation seen by many physiologists to limit severely the study of evolutionary physiology. The absence of the physiological fossil record is particularly frustrating as we see paleontology making major contributions to evolutionary morphology (Jablonski et al. 1986). Can the morphological fossil record help us to assess physiological evolution?

Certainly some spectacular (and hotly debated) assertions have been made about dinosaur physiology based on bone structure, among other evidence. Such assertions are relatively "safe" in that they are not tested against the physiology of living animals, but rather of extinct ones. Interestingly, our practical abilities to predict the physiology of living animals *solely* on the basis of morphological fossils have been tested several times, and have generally proven unreliable. Perhaps the best example concerns the coelacanths (Actinistia), a lineage well known from fossils. There was great optimism that the 1938 discovery of *Latimeria chalumnae* Smith would reveal much about the ancestors of land vertebrates. Fifty years later, however, *Latimeria* has not only clouded previously "clear" relationships among sarcopterygians (Forey 1980, 1988), but has also revealed many physiological surprises completely unanticipated from the extensive fossil record. For example, we failed to foresee *Latimeria*'s hypophyseal anatomy or urea-retaining osmoregulatory physiology (Lagios 1979), sluggish respiratory capacity (Hughes 1980), electrosensory capabilities (Bemis and Hetherington 1982), or even the locomotor uses of its paired fins (Fricke et al. 1987).

Even though the physiological fossil record is absent and we have argued above that the morphological fossil record is of limited use in predicting physiological features, evolutionary physiologists must not ignore phylogenetic reconstruction based on the morphological fossil record. Cladists who argue that phylogenetic patterns are more important than evolutionary

processes also deny that fossils play an important role in defining relationships among extant taxa (Patterson 1981; Forey 1982). This position leads to
perhaps an overreliance on physiologically significant characters in phylogeny.
The clearest example of this is Gardiner's (1982) cladistic interpretation of
sister-group relationships between birds and mammals ("Haemothermia"),
based in part on the close physiological and anatomical similarity between
mammalian and avian circulatory systems and the presence of homeothermy
(fig. 3). Opposing this view is a study confirming that knowledge of fossils
can be essential for phylogenetic reconstruction; we get a very different picture of relationships when fossils are ignored (Gauthier et al. 1988). These
workers used a large data set and computer-assisted cladistic methods
(PAUP) to test the impact of fossil synapsids on our interpretation of amniote phylogeny. In the absence of the fossils, "Haemothermia" prevails. But
as more and more synapsids are included in the analysis, the most parsimonious cladogram "flips" and leads to the traditional interpretation that birds
and mammals are in fact *not* sister groups. For lineages with a poor fossil
record, we can only speculate how much impact fossils might have had on
our perspective of relationships. But neither the position of the "pattern cladists" nor the poor quality of the record of some lineages justifies ignorance
of fossils. It seems desirable to use fossils not only for systematics but also
for inspiration: witness the many evolutionary morphologists who are stimu-
lated to pursue studies of living organisms by their analyses of fossils. Fossils
have much to teach us about evolutionary physiology, and we ignore them to
our detriment.

A final comment on the lack of a physiological fossil record concerns
the cloning of extinct genes. Recent experiments have successfully "resur-
rected" and cloned genes from "paleo-DNA" – genetic material still viable
due to unusual conditions of preservation (El-Mallakh 1987). Though cloned
paleo-DNA is a long way from a physiological process at the tissue level,
these recent events hold some degree of future promise for evolutionary
physiology.

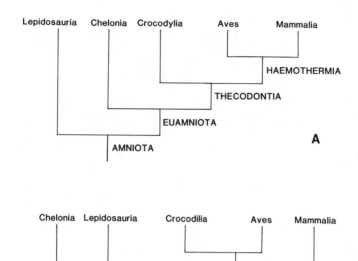

Figure 3. The influence of incorporating the fossil record into cladograms generated from physiological and anatomical features of living vertebrates based on Gauthier et al. (1988). (A) Gardiner's (1982) interpretation of sister-group relationships between birds and mammals ("Haemothermia"). (B) When information on synapsid fossils is incorporated, the most parsimonious cladogram shows mammals as distantly related to birds, and birds and crocodilia as sister groups. See text for discussion.

**Large Physiological Variability.** Another complicating factor in the study of physiological evolution is that physiological characters often inherently show more intraspecific variability than do the gross morphological structures that support them. For example, variation in heart rate in mice varies much more than does the ventricular mass of the heart, while at the gross anatomical level the hearts of individuals are indistinguishable (fig. 4) (Hou and Burggren, unpublished). In humans, the champion marathon runner may be very similar at the gross anatomical level to another person without the physiological stamina to run more than a few miles! This great variability in physiological characters complicates the study of physiological evolution be-

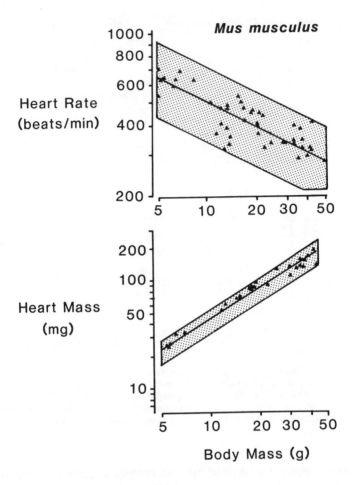

Figure 4. Variation in physiological function is often larger than variation in the anatomical structures from which function emanates. Data indicate the relationships between resting heart rate, heart mass, and body mass in domestic mice. The shaded region encloses all data points for each variable, while the straight line within the shaded region is a linear regression of the data points. Hou and Burggren, unpublished.

cause increased numbers of observations are required to establish central tendencies. While comparative physiologists have made an art of avoiding the study of variation, such heritable variation nonetheless is the source for evolutionary changes in physiology as well as for all other types of characters (Bennett 1987). It is apparent that the richness in variability of physiological traits has yet to be widely exploited in the study of microevolutionary physiology.

**Confusing Acclimation with Adaptation.** All animals respond physiologically to environmental change in their lifetimes, a process called *acclimation* (see Prosser 1986). Physiological characters of a species can change in evolutionary time, the process called *adaptation*. Much of the intraspecific physiological variation encountered by physiologists is a result of short-term physiological acclimation to short-term changes in environmental conditions. The time course of these changes in physiological variables can be amazingly short. For example, the blood of mice acclimated to a specific photoperiod regime shows a regular pattern of hourly changes in concentration of 2,3-DPG, an organophosphate compound that alters hemoglobin-oxygen binding (fig. 5, top panel) (Hutchison and Hazard 1981).

Perhaps because physiological acclimation can occur so rapidly, some biologists deliberately pass over physiological characters in phylogenetic analyses. They appear to assume (erroneously) that heritable physiological characters subject to rapid acclimation will also be subject to rapid adaptation during evolution, thus rendering the study of these characters useless in systematics and difficult to study from a microevolutionary perspective. This unnecessarily equates rates of acclimation with rates of adaptation. To illustrate this point, we cite the example of the control of hemoglobin-oxygen affinity of blood. While organophosphate modulators may vary within individuals due to acclimation, the "lifelong mean" concentration of such compounds is a heritable trait so stable in the evolutionary history of vertebrates as to establish separate subgroups within carnivores and artiodactyls on the basis of 2,3-DPG concentration (fig. 5). As another example, large minute-to-minute variations can occur in the preferred body temperature of individual lizards in thermal gradients, but the *average* preferred body temperature is a stable, predictable characteristic for all species within a genus (Huey and Bennett 1987).

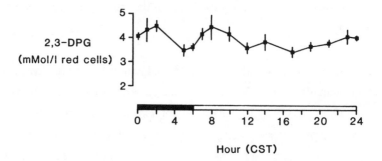

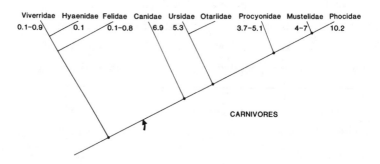

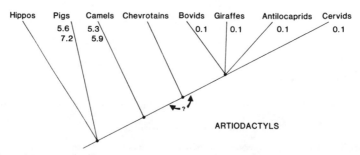

Figure 5. Short-term adjustments in physiological variables through acclimation do not necessarily indicate rapid evolutionary changes. Top panel: levels of 2,3-DPG within the blood of
the domestic mouse, *Mus musculus*, exposed to a 24-hour cycle of 6 hours darkness (black bar)
and 18 hours light. Changes in levels of 2,3-DPG are closely associated with changes in
hemoglobin-oxygen binding and blood-gas transport. From Hutchison and Hazard 1981.
Middle and bottom panels: cladograms of carnivores (middle) and artiodactyls (bottom), with
concentrations of 2,3-DPG (mmoles/l red blood cells) indicated for groups with available data.
The arrows indicate the point of divergence with respect to 2,3-DPG concentration. Modified
from Bunn 1980.

Finally, we emphasize that the degree to which an animal can acclimate physiologically is almost certainly heritable (Levins 1979; Koehn 1987; Powers 1987; Tsuji 1988). Indeed, the ability to acclimate might be evolutionarily conservative, while the manifestations of this trait – i.e., apparent great physiological variability in an organism – may be extremely labile within an individual. As long as identifiable traits are heritable, we ought to study them in the context of evolutionary physiology and the broader context of evolutionary biology.

**Excessive Reductionism.** Many physiologists remain "reducto-adaptationist" in perspective – subdividing organisms into multiple components and devising adaptive scenarios to explain organic diversity (Gould and Lewontin 1978). Of course, constructing scenarios on *organismal* evolution is very important (Gans 1985; Greene 1985; Bennett 1987), but most organ-system physiologists still focus on reductionism (see Ross 1981; Burggren 1987a; Feder 1987).

Why is this reductionist approach, often spurning investigation of variation within species or meaningful phyletic comparisons, so persistent? Certainly, reductionism has yielded enormous dividends. For example, early insight into nerve cell function stemmed largely from the study of giant squid axons. Would this field have advanced so rapidly if there had been heavy emphasis on cladistic comparison of general properties of nervous systems across cephalopods or even Mollusca? Obviously not. But all too often comparative physiological studies from their inception have little applicability to evolutionary questions because of the taxa studied (Huey 1987). As Bennett (1987) discusses in detail, animals are usually chosen for comparative physiological experimentation either on the basis of extreme physiological characters or because the animal is conducive to a certain physiological technique (i.e., squid axons 1 mm in diameter can be punctured relatively easily by microelectrodes). This manner of choosing animals, known as the Krogh principle (Krebs 1975), is not concerned with whether a species occupies a key position (or *any* position!) within a putative evolutionary sequence.

Although the Krogh principle has strongly influenced at least two generations of comparative physiologists, its original intent – to help understand how animals have become adapted to their environments – has frequently been corrupted in casual attempts to understand the evolution of physiological characters.

**Typology.** To answer fundamental physiological questions, comparative physiologists often turn to conveniently available "model" species (e.g., *Rana pipiens; Salmo gairdneri; Dipsosaurus dorsalis; Manduca sexta*).[2] An unfortunate consequence of the heavy use of model species is that typology dominates most initial attempts to study physiological evolution. That these so-called model species represent only a small fraction of the diversity of their lineage and are often quite atypical species does not seem to have been recognized, or has been largely ignored, by most comparative physiologists. Yet, the use of "the cockroach as insect," "the frog as amphibian," or "the turtle as reptile" persists, in spite of clear evidence of the dangers of this approach (Gans 1970). Not surprisingly, this type of comparative physiology has neither contributed much to evolutionary theories nor drawn upon them to formulate and test hypotheses in evolutionary physiology.

**Adaptationism.** Despite the best efforts of Gould and Lewontin (1978), Bartholomew (1987), and other critics of optimization theory in biology, the adaptationist program prevails in many physiological studies (e.g., Taylor and Weibel 1981). Perhaps the sense that many physiological parameters can be simultaneously optimized is a spin-off of the great variability of certain physiological characters, which gives us the misleading impression that organisms can greatly vary and finely tune these features during evolution. We must remind ourselves that organisms are not perfectly designed (Bartholomew 1987) and that the functions we study reflect phylogenetic as well as equilibrium limitations and compromises.

## Tools to Meet the Challenge of Evolutionary Physiology

Having discussed possible pitfalls in evolutionary physiology, we turn now to methods for its study. We draw upon tools used in functional morphology because both of these disciplines are *process-oriented* approaches to organismal biology. When physiologists or functional morphologists turn to evolutionary questions, the same general principles ought to apply.

**Phylogenetic Analysis.** *Thinking Systematically.* So much has been written about modern systematics and the need and methods for making rigorous phylogenetic comparisons (Ghiselin 1984; Patterson 1987; Huey 1987; Huey and Bennett 1987; Burggren, in press) that there is no need to reanalyze these debates. But one general lesson can be learned from the systematic upheavals in the last three decades: it is often inappropriate to adopt intact a "convenient" phylogenetic interpretation from the literature. In consequence of the revolution in systematics *all biologists must accept responsibility for the phylogenetic hypotheses they use.* The debate between cladistics and phenetics has produced one of the goals which both schools originally (ostensibly) desired: a "deregulation" of systematics and reduction of authoritarianism. We know only a little about relationships among some of the most celebrated lineages (e.g., lungfishes, *Latimeria*, rhipidistians and tetrapods), and as evolutionary physiologists we must learn to see the lack of information as exciting and be willing to explore and to test alternative phylogenies.

Huey (1987) has provided a very useful rule of thumb for comparative physiologists: "compare the set of closest relatives that show adequate diversity for a given problem." However, recognizing the closest relatives requires systematic analysis, particularly if the focus is on qualitative (i.e., innovative) evolutionary changes. We suggest an extension to Huey's rule of thumb: it is essential to identify and to compare closely the first out-group which *does not* have the feature in question. (See also Maddison et al. 1984.)

Systematics is unpopular in some circles, perhaps because it is "boring," "old-fashioned," or "too polemic." Nevertheless, systematics is the heart of biology and of evolutionary studies in particular (Wilson 1985). By analogy with recent progress in evolutionary morphology, a primary emphasis in evolutionary physiology must be on systematics, particularly on rigorous phylogenetic inferences presented as branching diagrams.

We refine and improve our evolutionary perspective by increasing our awareness of organismal diversity and systematics, but perhaps the major benefit of "thinking systematically" is that phylogenetic hypotheses can guide our comparisons and provide a logical basis for future studies. Testing phylogenetic predictions also offers one important kind of a priori theory against which we test our observations (Futuyma 1987). Placing physiological information on a branching diagram makes explicit the level of generality that a character is expected to have. For instance, it has recently been demonstrated that insulin-related peptides are produced by mollusks and that they may regulate shell growth (Smit et al. 1988). The insulin gene family thus has an earlier phylogenetic origin than previously suspected (fig. 6), allowing us to predict what groups may be expected to produce insulin and – as important from the viewpoint of directing future work – indicating where to look to find a lineage without the gene for insulinlike peptides.

There are several secondary benefits of thinking systematically. By focusing on species and their relationships, we are forced to return to the organismal level (Feder 1987), and by planning in advance to consider alternative phylogenetic arrangements, we avoid the pitfalls of basing our phylogenetic interpretations on single characters or single analytical techniques (e.g., maternal mitochondrial DNA). Finally, by making comparisons in the context of explicit phylogenetic diagrams we avoid "implicit" comparisons: too often, an older phylogenetic hypothesis (perhaps half-remembered from textbooks) guides our ideas of what comparisons ought to be most interesting. By getting the phylogeny out of our heads and onto paper where it can

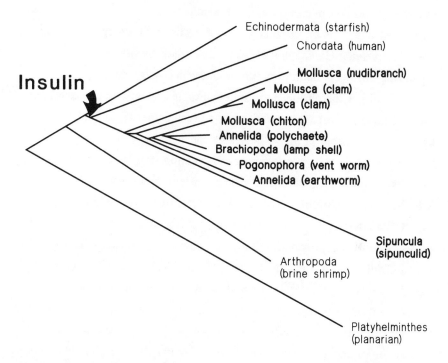

Figure 6. This dendrogram of selected invertebrates based on 18*S* rRNA sequences, from Field et al. 1988, shows how a new phylogeny can guide future study. The discovery of the insulin gene in mollusks (Smit et al. 1988) indicates that brachiopods, pogonophores, annelids, and sipunculids should be examined for the presence of insulin, while its absence should be confirmed in echinoderms and arthropods.

be criticized and further explored, we greatly advance the power of our generalizations.

*Strategies of Comparative Analysis.* Northcutt (1986) succinctly outlined the steps that any comparative analysis must follow. While he intended these for studies in comparative neurobiology, they apply equally well to physiological studies: (1) examine characters in a number of taxa; (2) recognize one or more patterns of character distribution; (3) establish the polarity of evolutionary change (i.e., ascertain primitive versus derived character states); (4) propose phylogenetic hypotheses based on the characters and their polarities; and (5) test the hypotheses and their corollaries through discovery of more characters, addition of new taxa, or reexamination of the characters. Here we suggest emphasis on the techniques of cladistics (Eldredge and Cracraft 1980; Wiley 1981; Coddington 1988) because they provide a clear means of organizing phylogenetic information in branching diagrams.

Cladistics, or phylogenetic systematics, is a codified approach to the study of homology that ostensibly seeks to reconstruct the actual branching pattern of the history of life. The search for these patterns does not rely on particular theories of evolutionary transformation (Patterson 1982), and because the resulting cladograms can be consistent with several different phylogenies, construction of a cladogram does not in theory yield a single phylogeny (Patterson 1987). However, most workers in evolution regard cladograms as a close approximation of actual phylogenetic history (Huey 1987).

Cladistics has carried the day philosophically, so much so that few would argue with the assertion that "all useful comparisons in biology depend on the relation of homology" (Patterson 1987). Clearly, however, there are alternatives, and "pure" cladistics may not always prevail in the practical business of generating and evaluating phylogenetic diagrams. Again, by analogy to evolutionary morphology, an open-minded approach to phylogenetic reconstruction has led to most of the recent progress (Liem and Wake 1985; Wake and Larson 1987). Nowhere are the benefits of open-mindedness more evident than in molecular systematics, the branch of systematics that has developed most recently and rapidly. Molecular systematists have used distance measures (i.e., phenetics), homology recognition (cladistics) and "hybrids" of these two approaches (McKenna 1987), and in the process have generated significant interest in systematics as well as new phylogenetic insights. In some cases, the explanations for noncladistic approaches seem to be of the "we don't know how it works, but it does work" variety. There is some truth to this: in the rare case where a phylogeny is known with certainty, as with the inbred mice studied by Fitch and Atchely (1987), molecular *distance* trees correlate exactly with the actual history of the lineages.

Physiologists by definition focus on functional characters. But should we base our phylogenetic hypotheses on physiological characters or attempt to "hang" those characters from phylogenies based on other characters? First, to avoid circular reasoning we should include as many uncorrelated characters as possible – morphological, physiological, biochemical, behavioral, or whatever – in generating phylogenetic hypotheses to test the evolution of functional characters. Second, there is recurring debate among systematists about the merits of basing phylogenies on characters whose functions are well understood (i.e., jaw-muscle morphology) or on those of unknown function (i.e., small differences in the pattern of skull roofing bones). To a large extent this argument seems moot. Clearly, the more we know about characters – their structure, function, and genetic basis – the more value they may have for phylogenetic analysis. It is particularly important to consider how features perceived as individual characters may be integrated into a single functional unit *which evolves as a unit.* Tight correlation of such physiological traits as high cardiac output, high gill or lung ventilation rates, and high blood-oxygen carrying capacity across taxa within a monophyletic lineage might indicate that they evolve as a character complex. Unfortunately, the concept of character complexes has only recently been emphasized by organ-system physiologists (e.g., Lindstedt and Jones 1987) and this seems to be an area in need of much additional study.

*Lessons from Modern Systematics.* Perhaps the most important practical lesson systematics can offer organ-system physiologists is how to recognize the problems of perspective and interpretation that follow directly from recognizing nonmonophyletic groups. Physiologists as a whole are among the worst consumers – abusers, if you will – of poorly diagnosed lineages. Paraphyly (the recognition and naming of a group that does not include all the descendants of a single common ancestor) is particularly problematic because it seems to be "only" a semantic argument. Yet the recognition of paraphyletic groups subtly influences the comparisons to be made, so that the comparisons may lose evolutionary meaning. Nowhere is this more obvious than in dealing with large, paraphyletic groups such as the "Reptilia" (fig. 7). As

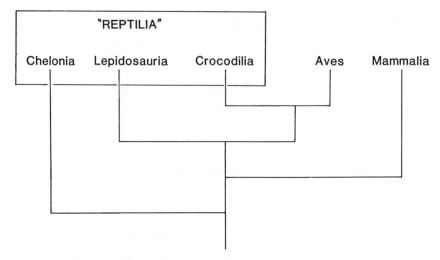

Figure 7. In this conventional phylogeny of the living amniotes, the "Reptilia" are a large paraphyletic grouping with very distantly related taxa.

long as we speak about reptiles as though they are a natural group, we perpetuate such mistakes of interpretation as the notion that crocodilians are more closely related to lizards than to birds. While it is true that many useful questions can be asked by gradal analysis (e.g., the concept of terrestrial ectothermic vertebrates, Pough 1980), it is important to realize that *gradal analyses alone cannot, in principle, answer evolutionary questions.* It is clear that such problems of perspective plague analyses of the physiological differences between lineages; even more significantly, they influence our choices of which physiological comparisons are most interesting. If we seek evolutionary explanations, we must focus on monophyletic groups.

A second lesson from recent progress in systematics is that parallelism and convergence are far more rampant than anticipated, and physiologists should be clearly aware of the consequences. While flight in bats and birds is easily recognized as convergent, other examples are less obvious. To give examples, air breathing in fishes has arisen independently several times (Randall et al. 1980; Little 1983). Lepidosirenid lungfishes have independently evolved a jaw-opening muscle that is remarkably similar to the depressor mandibulae of amphibians (Bemis 1988). No matter how interesting it is biologically, convergence offers few systematic inferences. For physiologists, the fact that convergence and parallelisms are so common reinforces the notion that we must be prepared to explore alternative phylogenetic interpretations. It also leads us now to discuss two approaches which temporarily suspend the attempt to generate and to test rigorously phylogenetic hypotheses.

**Feasibility Analysis.** The preceding section has highlighted approaches for studying evolutionary physiology through homology recognition and identification of out-groups which lack the character in question. This section relates to additional information from convergence and ontogeny available to determine the feasibility of an evolutionary scenario, the approach referred to as *transformationism* by Gould (1980). The evidence in these cases is circumstantial; it does not "stand alone" and is not directed at providing or testing phylogenetic schemes. Rather, its utility is in providing corroborating evidence for evolutionary interpretations developed from other characters.

*Convergence: Temporarily De-emphasizing Phylogeny.* Convergence is interesting because it shows clearly that adaptive evolutionary change occurs. (The functionalist's interest in convergence may be contrasted to the attitude of the "pure" systematist for whom convergence is only noise in the signal!) It is, of course, impossible to distinguish convergence from homology without a corroborated phylogenetic hypothesis. Once we have such a hypothesis, we can explore the convergently derived features by temporarily shifting our goals from phylogenetic pattern and asking *what evolutionary transformations are most feasible.* Feasibility analysis repeatedly reveals that forms not phylogenetically intermediate in a cladistic sense may show functionally intermediate states. It is thus useful to examine total diversity across broad taxa. If an intermediate condition necessary to our scheme is identified in a living animal – regardless of the phylogenetic position of that animal – then at the very least we know that *such a condition is feasible.* Systems that have evolved through convergent evolution – i.e., forms "proven" to work – suggest that comparable systems might also have evolved in unknown ancestral forms. In these cases, we need not invent a completely unknown ancestral physiology to bridge the gaps.

As an example of how useful this paradigm can be, consider the modeling of evolutionary transitions among amniotes from a single-pressure ventricle (the inferred primitive condition) to the dual-pressure ventricles of birds and mammals. Importantly, the feasibility paradigm will only provide insights on how this particular physiological character *might* have evolved, not how it *did* evolve! Living turtles, squamates and crocodilians are rich in variant forms of cardiovascular physiology and anatomy (Burggren 1987b). A surprisingly smooth, logical, and complete *physiological* transition can be generated using patterns of cardiovascular morphology and physiology of extant chelonians, squamates, crocodilians, and birds (fig. 8). In most chelonians and squamates the single ventricle, though complexly divided into separate *cava*, does not show complete anatomical separation during contraction of the heart. Consequently, the ventricle generates a uniform pressure throughout as it contracts and ejects blood into the arterial arches and thus functions as a single-pressure pump (Shelton and Burggren 1976). The ventricle of extant chelonians and squamates clearly represents a highly derived structure that bears little resemblance to that of birds (or mammals).

Interestingly, within squamates there is an excellent example of an intermediate form in our hypothetical transition. The heart of varanid lizards (*Varanus*) resembles that of other squamates, with the important exception that the most centrally located of the three ventricular cavities (the *cavum venosum*) has become much smaller (fig. 8). This morphological change appears pivotal in cardiac function, for it allows, during ventricular contraction, a brief anatomical separation of the cavity perfusing the lungs from the cavity perfusing the systemic circulation. During this period of anatomical separation, the ventricle functions as two distinct pumps, the left (*cavum arteriosum*) producing high pressures in the systemic arteries and the right (*cavum*

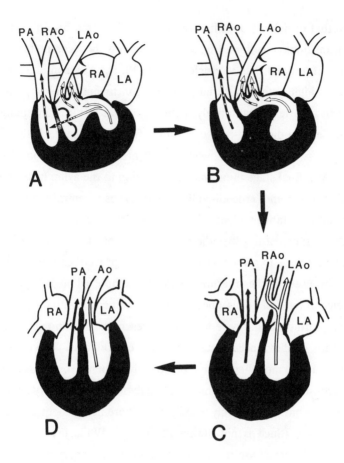

Figure 8. A "feasibility study" of heart evolution in amniotes from the single-pressure ventricle to the dual-pressure ventricle. The heart structure and function in a series of extant animals form a smooth hypothetical transition: (A) chelonian, (B) varanid lizard, (C) crocodilian, (D) bird. PA, pulmonary artery; LAo, Left Aorta; RAo, right aorta; RA, right atrium; LA, left atrium. (A) and (B) modified from Heisler and Glass 1985.

*pulmonale*) producing low pressure in the pulmonary arteries (Burggren and Johansen 1982). Thus, a comparatively minor morphological change produces an extremely important quantitative change in cardiac performance.

Continuing our temporary suspension of phylogeny, further diminishment of the *cavum venosum* and the fusion of free muscular ridges against opposing walls of the ventricle of a varanidlike heart could yield a heart very similar to that of living crocodilian reptiles, which represents an example of the next transitional condition (fig. 8). Crocodilians show complete anatomical division of left and right ventricles (White 1976). As a consequence, they can generate high systemic arterial pressures concurrent with low pulmonary pressures in a pattern qualitatively identical to that in birds and mammals. The major feature distinguishing crocodilian from avian hearts is that crocodilians retain two aortic arches, with the left arch emanating from the right ventricle. As a final stage in our model transition to the avian heart, the loss of the left aorta from a crocodilianlike heart leaves a central cardiovascular arrangement qualitatively identical to that of birds (fig. 8).

The purpose of this hypothetical modeling of the evolution of a dual-pressure ventricle is to show that a few existing "proven" cardiovascular systems can be used as the intermediaries in a simple (though hypothetical) physiological transition. Thus, ancestral amniote lineages with turtlelike, then varanidlike, cardiovascular systems could represent the evolutionary transition leading to the cardiovascular system of extant crocodilians. The main utility of this approach to modeling, which temporarily suspends our focus on phylogenetic reconstruction, lies in demonstrating *feasible* evolutionary transitions. Of course, feasibility studies rely heavily upon there being adequate variation in living forms. (Such is not always the case.) While users of such approaches should never aspire to reveal actual evolutionary pathways, such studies combined with phylogenetic analysis are an important component of putative evolutionary scenarios.

*Ontogeny: Totally Ignoring Phylogeny.* Ontogeny, of special interest to evolutionary physiologists, provides a second approach to feasibility analysis. Evolutionists have historically abused ontogeny as a probe of phylogeny (Gould 1977), and we must be careful eventually to place our developmental information in a rigorous phylogenetic context (Bemis 1984).

From a physiological perspective, analysis of developmental changes in living animals may provide insights into scenario generation *if we avoid overzealous interpretations.* Ontogenetic studies have proven particularly use-

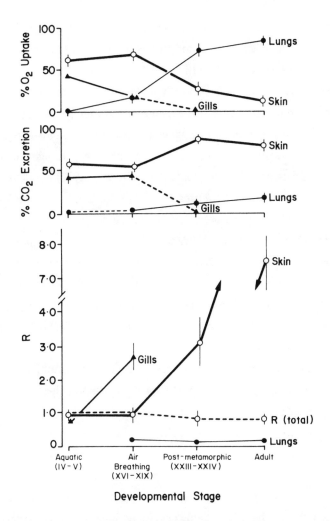

Figure 9. Partitioning of $O_2$ uptake and $CO_2$ elimination between skin, lungs, and gills in a series of developmental stages of the bullfrog, *Rana catesbeiana*. R is the gas exchange ratio, $CO_2$ elimination/$O_2$ uptake. Developmental stages use the Taylor-Kollros system. From Burggren and West 1982.

ful when developmental stages of a particular organism reflect significant changes in habitat or niche. For example, in many amphibians development culminating in metamorphosis results in dramatic physiological as well as morphological changes. To the evolutionary physiologist, amphibian development represents a model of the evolution of terrestriality that provides a proven system in which a transition from aquatic to air breathing actually occurs and can be studied in depth. In a sense, this makes some of the same assumptions and reaps the same dividends as the paradigm incorporating convergence outlined above. An additional advantage of this approach is that, initially, it requires no phylogenetic assumptions whatsoever.

To demonstrate the use of the ontogenetic paradigm, consider the evolution of skin breathing in amphibians. Romer (1972), on the basis of the fossil record, argued that skin breathing in extant amphibians evolved secondarily, and was not a characteristic of early amphibians. He believed that early amphibians, which had "advanced" lung ventilatory mechanisms and in many cases retained functional gills, were able to excrete sufficient $CO_2$ in these ways, a development correlated with the growth of heavy scales covering the body surface. On the other hand, he believed that living amphibians, with degenerate ventilation mechanisms and no gills, required naked skin to effectively eliminate $CO_2$.

Recent studies of gas exchange in developing amphibians have provided an opportunity to test certain aspects of Romer's hypothesis. For example, Romer (1972) suggested that possessing well-developed gills can alleviate the need for cutaneous gas exchange. Yet, in a variety of anuran larvae with highly developed internal gills and a sophisticated regulatory mechanism for matching gill ventilation to environmental conditions, the skin nonetheless is the major route for $CO_2$ elimination (fig. 9). It is implied in Romer's (1972) arguments that more highly developed lungs with more effective ventilatory mechanisms can provide an adequate route for $CO_2$ elimination. The pattern of change of gas-exchange partitioning between gills, lungs, and skin during development in individual species of extant anurans does not support this view. In ranid frogs, late larval development culminating in metamorphosis is accompanied by an increase in lung surface area and air breathing frequency (see Burggren 1988 for review). As the gills degenerate, the lungs are capable of assuming the great majority of $O_2$ uptake. Yet, even though the lungs have proliferated greatly during ontogeny, they have not assumed a much-enhanced role in $CO_2$ elimination, which continues primarily across the skin. In numerous amphibian species, cutaneous elimination of $CO_2$ does not emerge as a consequence of underdeveloped pulmonary function.

Thus, several of Romer's (1972) interpretations, based on the fossil record, of physiological transitions from water to land are not supported by physiological data from living animals that *have actually gone through the process.* Once again, this is only a feasibility study of physiological evolution using ontogeny. Our intention in this case study is not to say that Romer's (1972) interpretation is incorrect, but *to show that it is feasible that alternative physiological transitions could have occurred.* The utility of such ontogenetic feasibility studies is in providing corroborative evidence for scenarios based on systematic analyses.

*The Role of Feasibility Studies in Evolutionary Physiology.* Like circumstantial evidence in a courtroom, circumstantial evidence in evolutionary physiology must be very convincing and abundant before it becomes credible. A single feasibility study using either convergence or ontogeny can show only what could have been. The power in such approaches comes when multiple studies are combined, and when a problem is looked at from the viewpoint of both convergence and ontogeny.

This is well illustrated by our current understanding of acid-base physiology. Aquatic animals have high ventilation rates because water contains little $O_2$ relative to air. As a consequence of high ventilation and the fact that water has a very high capacitance for $CO_2$, aquatic animals retain very little molecular $CO_2$ within their body fluids. Terrestrial animals have relatively low ventilation rates because $O_2$ is so abundant in air. As a result, air-breathing animals retain much more $CO_2$ within their body fluids. They regulate acid-base balance by adjusting ventilation rates, and thus change the amount of retained $CO_2$ and the resulting body-fluid pH. Aquatic animals are unable to widely adjust body $CO_2$ levels because this molecule cannot be retained in the face of the unavoidably high ventilation rates. Thus, water-breathing animals use active transport of $HCO_3^-$ and $H^+$ across the body surfaces to regulate acid-base requirements. Importantly, a transition between these two mechanisms for acid-base regulation has been observed repeatedly in comparisons of water-breathing, amphibious, and air-breathing animals. Additionally, convergence feasibility studies have shown that this pattern holds in many studies looking at hypothetical transitions in fishes, amphibians, decapod crustacea, and mollusks. Ontogenetic studies drawn from amphibians, reptiles, birds, and mammals also indicate a transition of acid-base physiology as water breathing is replaced by air breathing during development. Assuming that the basic chemistry of carbon dioxide in water-based solvents has not changed in the last 300 million years (an entirely reasonable assumption), then we can be fairly confident about how ancestral aquatic animals, both invertebrate and vertebrate, regulated acid-base balance, and how these processes evolved with the multiple independent appearances of terrestrial animals. Finally, now that we understand something of how acid-base physiology may have evolved, we can begin to draw limits around other physiological and anatomical processes. For example, the transfer of acid-base regulation to gas-exchange organs as terrestriality evolved placed new demands and constraints on kidney structure and function.

Feasibility studies in evolutionary physiology can help answer broader questions in evolutionary biology. Evolutionary studies of specific systems may reveal the limits to other structures and functions. What we hope to accomplish is analogous to defining the possible "adaptive space" available to organisms (as in Raup and Michelson's 1965 analysis of shell morphology in mollusks).

### The Future and Promise of Evolutionary Physiology

Historically, physiological studies have contributed little to evolutionary biology. In part, this is because of tangible difficulties in studying physiological evolution (lack of fossil record, loose linkage of structure and function, large physiological variability, etc.) and the inability or unwillingness of comparative physiologists to adapt modern evolutionary approaches in favor of "reducto-adaptationist" methodologies. The former set of problems, though real, can be solved by various approaches. In this essay we have indicated several methodologies and paradigms (many of which are common in evolutionary morphology) that can be modified for the study of evolutionary physiology.

The latter set of problems, essentially concerning how physiologists go about their science, is in many ways the more difficult of the two. We certainly do not wish to imply that all comparative physiologists must put their studies in an evolutionary context. However, those that choose to should design, execute, and interpret their studies using modern, rigorous methodologies in evolutionary biology (Burggren, in press). As has been emphasized by ecological physiologists working from molecules to populations (Feder et al. 1987), to continue simply to categorize new physiological adaptations in additional animal species – i.e., to contribute in an encyclopedic fashion to the existing large comparative physiology data base – will make little additional contribution to any future role that comparative physiology may play in evolutionary biology.

There is no reason a priori why the study of physiology should not make crucial contributions to the study of evolution. Physiological processes, as cornerstones of life, are as much subject to macro- and microevolutionary processes as morphology, behavior, or biochemical processes. Indeed, because physiological processes are so adjustable in the short term with little associated anatomical change, it might emerge that physiological change is at the heart of rapid evolutionary change. A similar assertion has been made for the evolutionary impact of complex behaviors (Sage et al. 1985, who suggest that social drive facilitates rapid evolution of cichlid fishes). The idea of a physiological drive in evolution has been almost completely ignored (but see Huey and Bennett 1987).

In conclusion, we believe that evolutionary physiology can, should, and will be part of contemporary evolutionary biology. We are certain that no systematist feels so confident of morphological data as to spurn physiological cladograms in broad cladistic analyses. The onus is on comparative physiologists to provide this information, and thus elevate evolutionary physiology to equal footing with evolutionary morphology.

## Acknowledgments

Dr. Raymond Huey provided many useful comments during the preparation of this manuscript. The authors were supported by grants from the National Science Foundation (WEB, WWB) and the Whitehall Foundation (WEB).

## Endnotes

1. Homology, like many often-defined and much-debated concepts, may require an explicit definition: "A character of two or more taxa is homologous if this character is found in the common ancestor of these two taxa, or, two characters . . . are homologues if one is directly . . . derived from the other" (Wiley 1981:121). The cladistic interpretation of homology is to equate homologies with synapomorphies, i.e., shared derived characters (Patterson 1982). See also Roth (1984, 1988) and Patterson (1987).

2. The problem of emphasizing convenient model species is, of course, not limited to comparative physiologists. As long ago as 1950, Beach challenged comparative physiologists to cease their single-minded emphasis on white rats. Also, the medical literature is so heavily infiltrated by the concept of "model spoecies" as to be completely beyond redemption!

# References

Arnold, S. J. 1987. Genetic correlation and the evolution of physiology, pp. 76-98. In *New Directions in Ecological Physiology*, ed. M. E. Feder, A. F. Bennett, W. W. Burggren, and R. B. Huey. New York: Cambridge University Press.

Bartholomew, G. A. 1987. Interspecific comparison as a tool for ecological physiologists, pp. 11-35. In *New Directions in Ecological Physiology*, ed. M. E. Feder, A. F. Bennett, W. W. Burggren, and R. B. Huey. New York: Cambridge University Press.

Beach, F. A. 1950. The snark was a boojum. *American Psychologist* 5: 115-24.

Bemis, W. E. 1984. Paedomorphosis and the evolution of the Dipnoi. *Paleobiology* 10: 293-310.

Bemis, W. E. 1988. Convergent evolution of jaw-opening muscles in lepidosirenid lungfishes and tetrapods. *Canadian Journal of Zoology* 65: 2814-17.

Bemis, W. E., and T. E. Hetherington. 1982. The rostral organ of *Latimeria chalumnae*: Morphological evidence of an electroreceptive function. *Copeia* 1982: 467-71.

Bennett, A. F. 1987. Interindividual variability: An underutilized resource, pp. 147-66. In *New Directions in Ecological Physiology*, ed. M. E. Feder, A. F. Bennett, W. W. Burggren, and R. B. Huey. New York: Cambridge University Press.

Bunn, H. F. 1980. Regulation of hemoglobin function in mammals. *American Zoologist* 20: 199-211.

Burggren, W. W. 1987a. Invasive and noninvasive methodologies in ecological physiology: A plea for integration, pp. 251-72. In *New Directions in Ecological Physiology*, ed. M. E. Feder, A. F. Bennett, W. W. Burggren, and R. B. Huey. New York: Cambridge University Press.

Burggren, W. W. 1987b. Form and function in reptilian circulations. *American Zoologist* 27: 5-19.

Burggren, W. W. 1988. Lung structure and function: Amphibians, pp. 153-92. In *Comparative Pulmonary Physiology: Current Concepts*, ed. S. C. Wood. New York: Dekker.

Burggren, W. W. 1989. Does comparative respiratory physiology have a role in evolutionary biology (and vice versa)? In *Comparative Insights into Strategies for Gas Exchange and Metabolism*, ed. A. J. Woakes, M. K. Grieshaber, and C. R. Bridges. Cambridge: Cambridge University Press.

Burggren, W. W., and K. Johansen. 1982. Ventricular hemodynamics in the monitor lizard, *Varanus exanthematicus*: Pulmonary and systemic pressure separation. *Journal of Experimental Biology* 96: 343-54.

Burggren, W. W., and N. H. West. 1982. Changing respiratory importance of the gills, skin and lungs during metamorphosis in the bullfrog, *Rana catesbeiana*. *Respiratory Physiology* 47: 77-78.

Charlesworth, B., R. Lande, and M. Slatkin. 1982. A neo-Darwinian commentary on macro-evolution. *Evolution* 36: 474-98.

Coddington, J. A. 1988. Cladistic tests of adaptational hypotheses. *Cladistics* 4: 3-22.

Davis, D. D. 1958. The proper goal of comparative anatomy. *Proceedings of the Centenary and Bicentenary Congress of Biology, Singapore*, pp. 44-50.

Eldredge, N., and J. Cracraft. 1980. *Phylogenetic Patterns and the Evolutionary Process*. New York: Columbia University Press.

El-Mallakh, R. S. 1987. Cloning extinct genes. *Cryptozoology* 6: 49-54.

Feder, M. E. 1984. Consequences of aerial respiration for amphibian larvae. *Perspectives in Vertebrate Science* 3:71-86.

Feder, M. E. 1987. The analysis of physiological diversity: The prospects for pattern documentation and the general questions in ecological physiology, pp. 347-51. In *New Directions in Ecological Physiology*, ed. M. E. Feder, A. F. Bennett, W. W. Burggren, and R. B. Huey. New York: Cambridge University Press.

Feder, M. E., A. F. Bennett, W. W. Burggren, and R. B. Huey, eds. 1987. *New Directions in Ecological Physiology*. New York: Cambridge University Press.

Felsenstein, J. 1985. Phylogenies and the comparative method. *American Naturalist* 125: 1-15.

Field, K. G., G. J. Olsen, D. J. Lane, S. J. Giovannoni, M. Y. Ghiselin, E. C. Raff, N. R. Pace, and R. A. Raff. 1988. Molecular phylogeny of the animal kingdom. *Science* 239: 748-53.

Fitch, W. M., and W. R. Atchley. 1987. Divergence in inbred strains of mice: A comparison of three different types of data, pp. 203-16. In *Molecules and Morphology in Evolution: Conflict or Compromise?* ed. C. Patterson. London: Cambridge University Press.

Forey, P. L. 1980. *Latimeria*: A paradoxical fish. *Proceedings of the Royal Society of London*, Ser. B 208: 369-84.

Forey, P. L. 1982. Neontological analysis versus paleontological stories, pp. 119-57. In *Problems of Phylogenetic Reconstruction*, ed. K. A. Joysey and A. E. Friday. London: Academic Press.

Forey, P. L. 1988. Golden jubilee for the coelacanth *Latimeria chalumnae*. *Nature* 336: 727-32.

Fricke, H., O. Reinicke, H. Hofre, and W. Nachtigall. 1987. Locomotion of the coelacanth *Latimeria chalumnae* in its natural environment. *Nature* 329: 331-33.

Futuyma, D. J. 1987. Interindividual comparisons: A discussion, pp. 240-47. In *New Directions in Ecological Physiology*, ed. M. E. Feder, A. F. Bennett, W. W. Burggren, and R. B. Huey. New York: Cambridge University Press.

Gans, C. 1970. Respiration in early tetrapods - the frog is a red herring. *Evolution* 24: 723-34.

Gans, C. 1985. Scenarios: Why? In *Evolutionary Biology of Primitive Fishes*, ed. R. E.

Foreman, A. Gorbman, J. M. Dodd, and R. Olsson. New York: Plenum Press.

Gardiner, B. 1982. Tetrapod classification. *Zoological Journal of the Linnean Society* 74: 207-32.

Gauthier, J., A. J. Kluge, and T. Rowe. 1988. Amniote phylogeny and the importance of fossils. *Cladistics* 4: 105-209.

Gee, J. H., and J. B. Graham. 1978. Respiratory and hydrostatic functions of the intestine of the catfishes *Hoplosternum thoracatum* and *Brochis splendens* (Callichthyidae). *Journal of Experimental Biology* 74: 1-16.

Ghiselin, M. T. 1984. Narrow approaches to phylogeny: A review of nine books on cladism, vol. 1, pp. 209-22. In *Oxford Surveys in Evolutionary Biology*, ed. R. Dawkins and M. Ridley. Oxford: Oxford University Press.

Gilbert, S. F. 1988. *Developmental Biology*. Sunderland, MA: Sinauer Associates.

Goodman, M., M. M. Miyamoto, and J. Czelusniak. 1987. Pattern and process in vertebrate phylogeny revealed by coevolution of molecules and morphologies, pp. 141-76. In *Molecules and Morphology in Evolution: Conflict or Compromise?* ed. C. Patterson. Cambridge: Cambridge University Press.

Goslow, G. E., K. P. Dial, and F. A. Jenkins. 1989. The avian shoulder: An experimental approach. *American Zoologist* 29: 287-301.

Gould, S. J. 1977. *Ontogeny and Phylogeny*. Cambridge: Harvard University Press.

Gould, S. J. 1980. Is a new and general theory of evolution emerging? *Paleobiology* 6: 119-30.

Gould, S. J., and R. C. Lewontin. 1978. The spandrels of San Marco and the Panglossian Paradigm: A critique of the adaptationist programme. *Proceedings of the Royal Society of London*, Ser. B. 205: 581-98.

Greene, H. W. 1985. Diet and arboriality in the emerald monitor, *Varanus prasinus*, with comments on the study of adaptation. *Fieldiana: Zoology*, N.S., 31: 1-12.

Hartnoll, R. G. 1988. Evolution, systematics and geographical distribution, pp. 6-54. In *Biology of the Land Crabs*, ed. W. W. Burggren and B. R. McMahon. New York: Cambridge University Press.

Heisler, N. H., and Glass, M. L. 1985. Mechanisms and regulation of central vascular shunts in reptiles, pp. 334-47. In *Cardiovascular Shunts: Phylogenetic, Ontogenetic and Clinical Aspects*, ed. K. Johansen and W. Burggren. Copenhagen: Munksgaard.

Herring, S. 1988. Introduction: How to do functional morphology. *American Zoologist* 28: 189-92.

Huehns, E. R., N. Dance, G. H. Beaven, F. Hecht, and A. G. Motulsky. 1964. Human embryonic hemoglobins. *Cold Spring Harbor Symposia on Quantitative Biology* 29: 327-33.

Huey, R. B. 1987. Phylogeny, history and the comparative method, pp. 76-101. In *New Directions in Ecological Physiology*, ed. M. E. Feder, A. F. Bennett, W. W. Burggren, and R. B. Huey. New York: Cambridge University Press.

Huey, R. B., and A. F. Bennett. 1987. Phylogenetic studies of coadaptation: Preferred temperatures versus optimal performance temperatures of lizards. *Evolution* 41: 1098-1115

Hughes, G. M. 1980. Ultrastructure and morphometry of the gills of *Latimeria chalumnae* and a comparison with the gills of associated fishes. *Proceedings of the Royal Society of London,* Ser. B 208: 309-28.

Hutchison, V. H., and E. S. Hazard. 1981. Diel variation of erythrocytic phosphates in *Mus musculus* acclimated to different photoperiod regimes. *Comparative Biochemistry and Physiology* 70A: 9-12.

Jablonski, D., S. J. Gould, and D. M. Raup. 1986. The nature of the fossil record: A biological perspective, pp. 7-22. In *Patterns in the History of Life. Dahlem-Konferenzen 1986,* ed. D. M. Raup and D. Jablonski. Berlin: Springer-Verlag.

Keller, R. E. 1981. An experimental analysis of the role of bottle cells and the deep marginal zone in the gastrulation of *Xenopus laevis. Journal of Experimental Zoology* 216: 81-101.

Koehn, R. K. 1987. The importance of genetics to physiological ecology, pp. 170-85. In *New Directions in Ecological Physiology,* ed. M. E. Feder, A. F. Bennett, W. W. Burggren, and R. B. Huey. New York: Cambridge University Press.

Krebs, H. A. 1975. The August Krogh Principle: "For many problems there is an animal on which it can be most conveniently studied." *Journal of Experimental Zoology* 194: 309-44.

Lagios, M. D. 1979. The coelacanth and chondrichthyes as sister groups: A review of shared apomorph characters and a cladistic analysis and reinterpretation. *Occasional Papers of the California Academy of Sciences* 134: 25-44.

Lagios, M. D., and J. E. McCosker. 1977. A cloacal excretory gland in the lungfish *Propterus. Copeia* 1977: 176-78.

Lannoo, M. J., and M. D. Backman. 1984. On flotation and air breathing in *Ambystoma tigrinum* larvae: Stimuli for and the relationship between these behaviors. *Canadian Journal of Zoology* 62: 15-18.

Lauder, G. V. 1982. Introduction, pp. xl-xlv. In *Form and Function,* ed. E. S. Russell. Chicago: The University of Chicago Press.

Levins, R. 1979. Coexistence in a variable environment. *American Naturalist* 114: 765-83.

Liem, K. F. 1978. Modulatory multiplicity in the functional repertoire of the feeding mechanism in cichlid fishes. *Journal of Morphology* 158: 323-60.

Liem, K. F., and D. B. Wake. 1985. Morphology: Current approaches and concepts, pp. 366-77. In *Functional Vertebrate Morphology,* ed. M. Hildebrand, D. M. Bramble, K. F. Liem, and D. B. Wake. Cambridge: Harvard University Press.

Lindstedt, S. L., and J. H. Jones. 1987. Symmorphosis: The concept of optimal design, pp. 289-305. In *New Directions in Ecological Physiology,* ed. M. E. Feder, A. F. Bennett, W. W. Burggren, and R. B. Huey. New York: Cambridge University Press.

Little, C. 1983. *The Colonisation of Land: Origins and Adaptations of Terrestrial Animals*. Cambridge: Cambridge University Press.

Maddison, W. P., M. J. Donoghue, and D. R. Maddison. 1984. Outgroup analysis and parsimony. *Systematic Zoology* 33: 83-103.

Maitland, D. P. 1986. Crabs that breathe air with their legs – *Scopimera* and *Dotilla*. *Nature* 319: 493-95.

Mayr, E., and W. B. Provine. 1980. *The Evolutionary Synthesis*. Cambridge: Harvard University Press.

McKenna, M. C. 1987. Molecular and morphological analysis of high-level mammalian interrelationships, pp. 55-93. In *Molecules and Morphology in Evolution: Conflict or Compromise?* ed. C. Patterson. London: Cambridge University Press.

McMahon, B. R., and W. W. Burggren. 1987. Respiratory physiology of intestinal air breathing in the teleost fish *Misgurnus anguillicaudatus*. *Journal of Experimental Biology* 133: 371-94.

Milsom, W. K. 1975. Development of buoyancy control in juvenile Atlantic loggerhead turtles, *Caretta c. caretta*. *Copeia* 1975: 758-62.

Northcutt, R. G. 1986. Strategies of comparison: How do we study brain evolution? *Verhandlungen der Deutsch Zoologische Gesellschaft* 79: 91-103.

Patterson, C. 1981. Significance of fossils in determining evolutionary relationships. *Annual Review of Ecology and Systematics* 12: 195-223.

Patterson, C. 1982. Morphological characters and homology, pp. 21-74. In *Problems of Phylogenetic Reconstruction*, ed. K. A. Joysey and A. E. Friday. London: Academic Press.

Patterson, C. 1987. Introduction, pp. 1-22. In *Molecules and Morphology in Evolution: Conflict or Compromise?* ed. C. Patterson. London: Cambridge University Press.

Payan, P., J. P. Girard and N. Mayer-Gostan. 1984. Branchial ion movements in teleosts: The roles of respiratory and chloride cells, pp. 39-64. In *Fish Physiology*, volume XB, ed. W. S. Hoar and D. J. Randall. New York: Academic Press.

Pough, F. H. 1980. The advantages of ectothermy for tetrapods. *American Naturalist* 115: 92-112.

Powers, D. A. 1987. A multidisciplinary approach to the study of genetic variation within species, pp. 102-30. In *New Directions in Ecological Physiology*, ed. M. E. Feder, A. F. Bennett, W. W. Burggren, and R. B. Huey. New York: Cambridge University Press.

Prosser, C. L. 1986. *Adaptational Biology*. New York: Wiley Intersciences.

Randall, D. J., W. W. Burggren, A. P. Farrell, and M. S. Haswell. 1980. *Evolution of Air Breathing Vertebrates*. New York: Cambridge University Press.

Raup, D. M., and A. Michelson. 1965. Theoretical morphology of the coiled shell. *Science* 147: 1294-95.

Romer, A. S. 1972. Skin breathing – primary or secondary? *Respiratory Physiology* 14: 183-92.

Ross, D. M. 1981. Illusion and reality in comparative physiology. *Canadian Journal of Zoology* 59: 2151-58.

Roth, V. L. 1984. On homology. *Biological Journal of the Linnean Society* 22: 13-29.

Roth, V. L. 1988. The biological basis of homology, pp. 1-26. In *Ontogeny and Systematics*, ed. C. J. Humphries. New York: Cambridge University Press.

Sage, R. D., P. V. Loiselle, P. Basasibwaki, and A. C. Wilson. 1985. Molecular versus morphological change among cichlid fishes (Pisces: Cichlidae) of Lake Victoria, pp. 185-220. In *Evolution of Fish Species Flocks*, ed. A. A. Echelle and I. L. Kornfield. Orono: University of Maine Press.

Scheid, P. 1982. A model for comparing gas-exchange systems in vertebrates, pp. 3-16. In *A Companion to Animal Physiology* ed. C. R. Taylor, K. Johansen and L. Bolis. New York: Cambridge University Press.

Schwartz, L., and W. Stühmer. 1984. Voltage-dependent sodium channels in an invertebrate striated muscle. *Science* 225: 523-25.

Shelton, G., and W. Burggren. 1976. Cardiovascular dynamics of the Chelonia during apnoea and lung ventilation. *Journal of Experimental Biology* 64: 323-43.

Smit, A. B., E. Vreugdenhil, R. H. M. Ebberink, W. P. M. Geraerts, J. Klootwijk, and J. Joosse. 1988. Growth-controlling molluscan neurons produce the precursor of an insulin-related peptide. *Nature* 331: 535-38.

Taylor, C. R., and E. R. Weibel. 1981. Design of the mammalian respiratory system. I. Problem and strategy. *Respiration Physiology* 44: 1-10.

Tsuji, J. S. 1988. Thermal acclimation of metabolism in *Sceloporus* lizards from different latitudes. *Physiological Zoology* 61: 241-53.

Vigna, S. R. 1985. Functional evolution of gastrointestinal hormones, pp. 401-12. In *Evolutionary Biology of Primitive Fishes*, ed. R. E. Foreman, A. Gorbman, J. M. Dodd, and R. Olsson. New York: Plenum Press.

Wake, D. B., and A. Larson. 1987. Multidimensional analysis of an evolving lineage. *Science* 238: 42-48.

Wallis, M. 1975. The molecular evolution of pituitary hormones. *Biological Reviews (Cambridge Philosophical Society)* 50: 35-98.

White, F. N. 1976. Circulation. In *Biology of the Reptilia*, ed. C. Gans. New York: Academic Press.

Wiley, E. O. 1981. *Phylogenetics: The Theory and Practice of Phylogenetic Systematics*. New York: John Wiley.

Wilson, E. O. 1985. Time to revive systematics. *Science* 230: 1227 (editorial).

# PALEONTOLOGY

# The Sand-Dollar Syndrome:
# A Polyphyletic Constructional Breakthrough

*Adolf Seilacher*

Innovation is a relative term. Evolutionary change implies innovative elements at any taxonomic level. As humans we would claim this status even for the individual. Nevertheless, in the present context, we mean major transformations that opened the door to a new ecologic zone in which a whole new group of organisms could evolve. Characteristically, such transformations require reorganization at several levels. Feathers, wings, toothless jaws, reduced tails and warm-bloodedness equally distinguish a bird from other sauromorphs. But since it is unlikely that all elements of the bird syndrome have evolved simultaneously, one may legitimately ask in every case for the key feature. In other words, we must structure the syndrome in a hierarchical fashion. In this approach, we may find that particular constructional requirements may be met by more than one solution. Such *alternative fabricational pathways* are the clues by which seemingly coherent adaptive groups can be phyletically disentangled.

## The Float of Scyphocrinites

Among the echinoderms, the pelagic crinoid *Scyphocrinites* may be cited as an illustrative example for the multiplicity of innovation, once the threshold to a new adaptive zone has been passed. Despite a short time range (uppermost Silurian), this crinoid has gained cosmopolitan distribution by a unique innovation: the transformation of the root into a gas-filled buoy (lobolith).

This organ allowed the animal to drift pseudoplanktonically at a time when drift logs were not yet available as an easier stepping stone (Seilacher et al. 1968). In a model study, Haude (1972) was able to show that the two species of *Scyphocrinites* have used radically different ways to construct and expand the calcareous walls of their loboliths. In *S. elegans* the root cirri continued to grow into a fractal system of branching spirals, which together formed a flexible tissue that could expand by continuous and coordinated growth of constituent ossicles and addition of new ones at the tips. In *S. stellatus* the terminal cirral ossicles stopped insertion of new elements. Instead they expanded into a sandwiched plate mosaic not unlike an echinoid or cystoid capsule. This case shows us how originally separated arms of an echinoderm can be transformed into a capsule. But since stereom fusion along the sutures has to be avoided in the interest of flexibility and further growth, the second solution also implies that adjacent ossicles must accommodate their crystallographic axes (mainly *a* and *b*) to different orientations not only with regard to their original series, but also to their new neighbors. The double pathway suggests that the establishment of a calcareous lobolith was preceded by a stage in which this organ was only a bladder of soft tissue spread between the root cirri.

We may hypothesize a still earlier stage in the evolutionary history of *Scyphocrinites* analogous to the present-day *Lepas fascicularis*. This gooseneck barnacle starts its life cycle in a conservative sessile fashion, but it becomes attached exclusively to the floats of siphonophores. Such an epiplanktonic mode of life, combined with miniaturization and protracted maturity (progenesis), was probably the evolutionary stepping stone into the pelagic realm. The present species, however, goes one step further by producing its own foam float as soon as it outgrows the siphonophoran starter substrate.

Assuming that the evolution of *Scyphocrinites* passed through similar steps, where should the innovation be placed: at the acquisition of the fully calcified and fossilizable lobolith, at its soft predecessor, or at the epiplanktonic stage, brought about by a change in larval settling behavior that marked the initial step into the pelagic realm?

Another general lesson told by the *Scyphocrinites* example is the rule of early experimentation, or rather the multiplicity of pathways, in which a basic invention is likely to be quickly optimized. A consideration of early cars and airplanes shows that this rule is not restricted to organic evolution. The taxonomic weighting of the alternative pathways towards optimized design is only a retrospective artifact. Had the promising clade of *Scyphocrinites* not been prematurely terminated by extinction, the difference between the two species would certainly have been granted the rank of different genera or families. At the same time, this example should remind us that the fate of clades during extinction events is not an adequate measure of their constructional fitness (Jablonski 1986). In the face of exceptional perturbations, survivorship appears to be determined by criteria other than those on which normal evolutionary selection concentrates. There is certainly as little reason to talk about poor design in *Scyphocrinites* as there is in ammonites relative to their luckier nautiloid relatives (Seilacher 1988).

## Sand Dollars

Compared to *Scyphocrinites,* sand dollars have the advantages that (1) the radiation of these echinoids was uninterrupted for at least sixty million years, (2) no intermediate stages of reduced fossilization potential are involved, and (3) many species still live in nearshore habitats, where we can directly observe the functions of distinctive features and apply them to fossil forms. But there is also a drawback. The basic innovation of *Scyphocrinites* – the switch from a benthic to a pelagic lifestyle – has hardly changed the mode of feeding. Sand dollars, in contrast, acquired a novel mode of nutrition compared to their clypeasteroid ancestors, while remaining in essentially the same habitat. Therefore, the point of departure cannot be sharply defined.

The natural history of sand dollars is well exemplified by the genus

*Mellita.* Its extant species inhabit shallow sandy bottoms in the Caribbean
and Panamanian provinces, cruising along a few millimeters below the sand
surface. This motion, for which the disc shape of the test is well suited, is
effected by relatively long locomotory spines on the flat lower (oral) surface
(fig. 1). The spines on the evenly vaulted upper (aboral) surface of the test
do not actively participate in locomotion. Their main task is to sieve from
a thin veneer of sand small food particles that fall below the canopy of
broadened spine tips and are then transported, by cilia and tube feet, around
the edge (or through the lunules) to the lower surface and mouth.

Without further detailing this process (Seilacher 1979; Ghiold 1979,
1984), we may state that sand dollars in the strict sense are clypeasteroid
echinoids that process the thin surface layer of the sand by gravitational sieving. In *Mellita,* the following features can be directly or indirectly referred to
this unique mode of feeding: (1) differentiation of locomotory spines on the
oral surface and of sieve spines on the aboral side, plus larger marginal
spines along the ambitus and lunular edges; (2) a disc-shaped test with a flat
base, evenly vaulted top, and sharp-edged ambitus; (3) sutural interlocking of
coronal plates by interpenetrating projections; (4) branched food-groove systems on the lower side; and (5) lunular perforations through the test.

Since none of these features are found in ancestral clypeasteroid families (or in other echinoids) but have convergently evolved in groups derived
from them, we may call this list of characters the "sand-dollar syndrome."

Further discussion will focus on each of these features with the following questions in mind: (1) Is there, for constructional or functional reasons,
a sequential order in which the features are likely to have evolved (temporal
hierarchy)? (2) Could an equivalent functional design be reached by more
than one morphogenetic mode (alternative morphogenetic pathways)?

**Spine Differentiation** (fig. 1). Echinoid spines are single ossicles growing
along the crystallographic *c* axis. They articulate with the test by ball joints,
but lack the ability to insert new elements at the growing tips. This con-

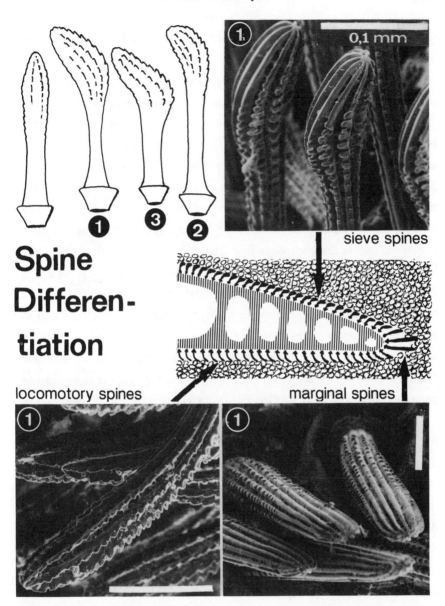

# Spine Differen- tiation

sieve spines

locomotory spines

marginal spines

Figure 1. The key feature in the sand-dollar syndrome is the differentiation of spines for specific functions. The canopy of club- or shoe-shaped spines on the upper (aboral) surface of the test constitutes a gravitational sieve for fine food particles, while the locomotory function falls solely to spines on the lower surface. Large, bladelike marginal spines around the ambitus and the lunules may perform steering and shielding functions. Circled numbers in this and following figures refer to the independent groups of sand dollars as shown in figure 8.

straint excludes not only their transformation into articulated legs, but also limits their potential morphospace.

In sand dollars, the relatively large marginal spines have a somewhat flattened, but otherwise normal, geometry. The slight bend of the locomotory spines is more unusual, but still allows accretionary growth in a fairly isometric fashion. Most distinctive are the sieve spines on the upper surface of the test because their tips are not only widened, but also meet the shaft at an angle ("shoe spines"). This shape is well suited to handle sand grains, particularly since serrate ridges provide the sole of the "shoe" with a gripping surface, but it cannot grow isometrically without resorption. Theoretically, one would expect these spines to first grow the slender shaft to the required length and only then add the geniculate club. Clubless miliary spines are in fact interspersed in the canopy (Ghiold 1984). Instead of a club, their tips bear a mucus balloon well suited to buffer the motion of the spines and to lubricate the passage of the sand. It is likely (but not yet proved) that mucus spines ultimately change their function by transforming into shoe spines. Once the shoe spines are ready, restriction of further growth does not impede their function, as long as the size of the club fits that of the sand grains.

Similar sieve spines are found in all sand dollars; they could be taken for a plesiomorph feature if we did not know from other criteria that in at least three groups (Scutellina, Rotulidae, and Arachnoididae) they must have evolved independently (Durham 1966; Kier 1982; Smith 1984). Unfortunately, the lack of alternative morphogenetic pathways in making a shoe spine renders this feature unsuitable for cladistic purposes. In addition, spines are rarely recorded in fossil species. Nevertheless, it is reasonable to assume that the differentiation of sieve spines represents the most basic element, and, therefore, the first step, in the sand-dollar syndrome.

**Test Flattening** (fig. 2). Echinoid tests in general can be understood as mineralized balloons, or "pneu" structures. This is shown not only by their overall geometries, but also by their descendence from flexible Paleozoic forms,

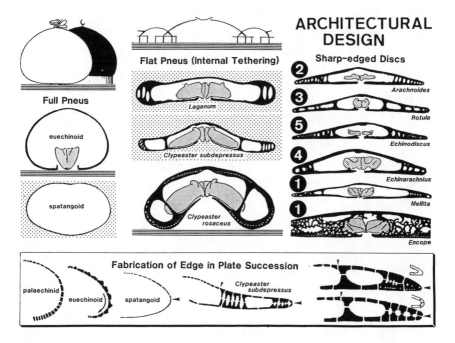

**ARCHITECTURAL DESIGN**

Full Pneus

euechinoid

spatangoid

Flat Pneus (Internal Tethering)

*Laganum*

*Clypeaster subdepressus*

*Clypeaster rosaceus*

Sharp–edged Discs

*Arachnoides*

*Rotula*

*Echinodiscus*

*Echinarachnius*

*Mellita*

*Encope*

Fabrication of Edge in Plate Succession

palaechinid

euechinoid

spatangoid

*Clypeaster subdepressus*

Figure 2. As a corollary of spine differentiation, sand-dollar tests become disc-shaped with a sharp edge, evenly vaulted roof, and flat lower surface. Internal pillar support and stopping of the ontogenetic conveyor belt motion (long arrows) around the edge, as reached already in pre-sand-dollar clypeasteroids, was a prerequisite for this extreme deviation from the balloon, or pneu, shape of echinoid tests. See text for further explanation. Modified from Seilacher 1979.

as well as by the teratological deviations of modern species from regular shapes in polluted habitats (Dafni 1986). In the Clypeasteroidea the dome shape has become flattened by internal tetherings across the body, the mineralized versions of which (with a growing transverse suture in the middle) function as supporting pillars. Rigid pillars, however, require that the meridional conveyor-belt motion of plates across the equator be stopped at an early ontogenetic stage. Although this step was a necessary prerequisite, we do not include it in the sand-dollar syndrome because it was initiated earlier in clypeasteroid history.

Sand-dollar tests are unique not only in a further depression of the height/width ratio. We also observe that the surfaces become more even:

on the oral side the sunken mouth opening is lowered so that the bottom side becomes perfectly flat, adequate for its sole locomotory responsibility. On the upper side of the test the petaloid depressions – and, in the Laganids, concentric depressions near the margins – are also flattened to produce the even vaulting of the sieve.

A still more distinctive feature, however, is the sharpening of the ambitus. This principal deviation from the pneu model can theoretically be effected in three ways: (1) the edge is established along the latitudinal sutures between successive plates in each row; (2) the edge is formed by angular edge plates; (3) these two modes alternate in adjacent plate series.

Recognition of these three modes in actual specimens is not easy because sutures are difficult to see without special preparation. An easier way would be the study of radial thin sections under polarized light. Since diagenesis alters the spongy stereom into solid calcite crystals, such studies would even be easier in fossil than in recent specimens. Unfortunately, such a survey is still pending. Yet it promises another line of evidence for polyphyletic origins, since the alternative solutions of the problem are equally valid but unlikely to be changed once the decision has been made for one or the other.

**Sutural Interlocking** (fig. 3). Once the transformation from flexibly plated balloons to rigid tests had been accomplished (a process that happened several times in echinoid phylogeny), echinoid plate sutures were held together by collagen fibers. This construction allowed the suture to open tensionally during growth periods. Given the angular movement of plates during the ontogenetic conveyor-belt motion around the ambitus, the calcitic stereom contributes to the bonding only by irregular knobby projections. This is still true for the sutures of clypeasteroid stock groups. Stopping of the conveyor belt in the sand dollars, however, lifted this constraint. In response, they improved the bonding by organized stereomic projections, but they did it differently in the three major groups: (1) the Arachnoidida organized the stere-

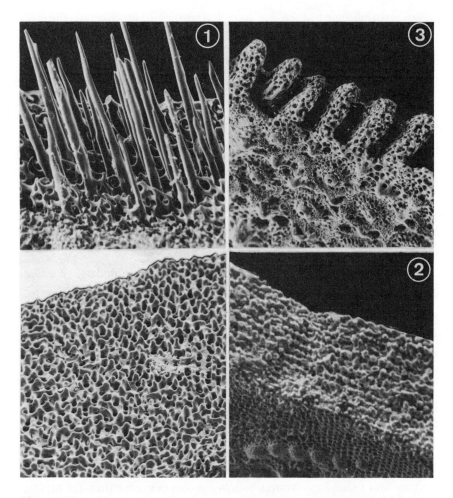

Figure 3. Sutural interlocking. Since conveyor-belt motion has stopped and crystallographic axes of the stereom differ in adjacent test plates, sand-dollar sutures may be strengthened by interlocking projections. While tubercles on the sutural surfaces of *Clypeaster* (lower left) lack a specific pattern, they form long nails in *Mellita* (upper left) and other members of the Scutellina (in the old sense). In contrast, tubercles are aligned parallel to the surface in the Arachnoidina (lower right), while the plates dovetail by massive stereomic projections in the Rotulina (upper right). Modified from Seilacher 1979.

omic knobs into rows parallel to the plate surface; (2) in the Scutellina the knobs grow into long needles that penetrate deeply into the stereom of the opposite plate; (3) the Rotulina, in contrast, dovetail the plates by heavy ridges perpendicular to the plate surface consisting of reticular stereoms.

Again, this element of the sand-dollar syndrome has not yet been systematically surveyed in all sand dollars. Nevertheless, it is clear that the fabricational alternatives provide a useful criterion for polyphyletic origins.

**Branched Food Grooves** (fig. 4). Food grooves, whether in medusae, crinoids, edrioasteroids or echinoids, are the way that microphagous animals efficiently transport food particles (often bound by mucus) from a large tributary area to the mouth. Since the mouth is centrally located, they approach the dendroid patterns of banana roads. In echinoderms the food transport is largely (though not exclusively) carried out by ambulacral feet; therefore, the grooves are centered in the ambulacral areas.

Food grooves are not a distinctive feature of sand dollars. We find them already in their clypeasteroid stem groups, but here they lack branching and run along the radial sutures from the ambitus straight to the mouth. It was not until the transition to sieve feeding that expansion into a branching banana-road system became necessary.

In the growing plate mosaic of an echinoid test, the dendroid pattern of such a road system must comply with the reticular one of the plate sutures. This fabricational constraint has been accommodated in different ways in various groups of sand dollars.

In the Rotulida and in all Scutellina (except for the Echinarachniidae) the ancestral radial food grooves bifurcate a short distance from the mouth, with more bifurcations following toward the periphery. Also, the branches have sharp kinks near the growth centers of the plates because the stretches between these centers are controlled by sutural growth.

In the Arachnoidida the primitive radial food grooves not only persist, but also extend around the ambitus to the upper surface. Their tributaries form a dense grid of featherlike groovelets, whose equidistant geometric pattern and diagonality with respect to the plate sutures suggest that they grow by coordinated steps, not unlike diagonal patterns in woven carpets. More

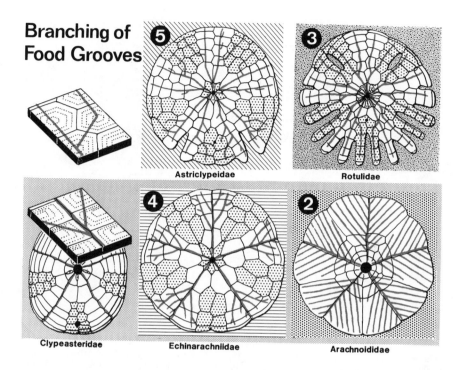

**Branching of Food Grooves**

Astriclypeidae

Rotulidae

Clypeasteridae

Echinarachniidae

Arachnoididae

Figure 4. Since sand dollars collect food on the aboral surface, but have their mouth on the lower side, banana-road systems of food grooves are another integral part of the sand-dollar syndrome. These food grooves are largely restricted to the oral side and to the ambulacral areas, but their branching patterns accommodate differently for plate growth. Members of Scutellina (including Astriclypeidae) and of the Rotulida have bifurcating systems, whose branches link the plate growth centers in an angular course. In contrast, Echinarachniidae and Arachnoididae maintain the median food groove of clypeasterid ancestors (note straightening of radial sutures!) and add branches cutting obliquely across sutures. Since the option for one of the two modes of branching is irrevocable, Echinarachniina are here considered as an independent group of sand dollars.

exactly, growth increments must equal the grid of the pattern so that each groovelet can link with the next one in the row on the opposite plate.

A completely different pattern is found in *Echinarachnius* and in the Eocene genus *Periarchus* (fig. 5), whose slightly raised apical system makes it resemble a Chinese hat rather than a shield. As in *Arachnoides,* the food-groove patterns of these genera consist of the ancestral radial groove plus diagonal side branches, but there are only very few ramifications, which are

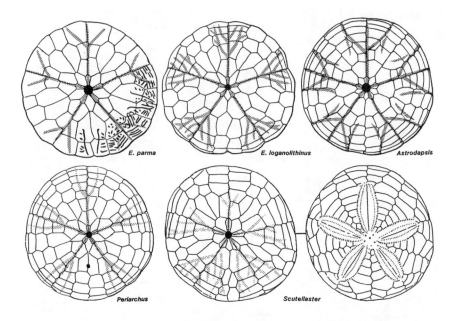

Figure 5. Echinarachniid food grooves. With the food groove pattern as a distinctive criterion, Echinarachniina can be traced back to the Eocene (*Periarchus*). Other fossil members include *Scutellaster*, whose eccentricity with regard to food grooves and apical system suggests a mode of life similar to *Dendraster* (upper right in fig. 7). Modified from *Treatise on Invertebrate Paleontology* and *Periarchus* specimen in Peabody Museum, Yale University.

restricted to the periphery of the oral surface. The way in which the side branches grow in this case is still problematic. In the diagonal comb patterns of *Arachnoides,* groovelets could change partners across the suture at every growth step. In the case of a single side branch, however, diagonality can only be maintained by lateral dislocation of the groove.

In summary, food-groove patterns confirm the polyphyletic picture derived from other criteria, if we grant that Scutellina and Rotulida coincidentally followed the same pathway. However, if we consider food-groove branching a primary feature of the sand-dollar syndrome, the different pattern in the Echinarachniidae suggests a fourth independent sand-dollar origin because one pattern cannot be secondarily transformed into the other.

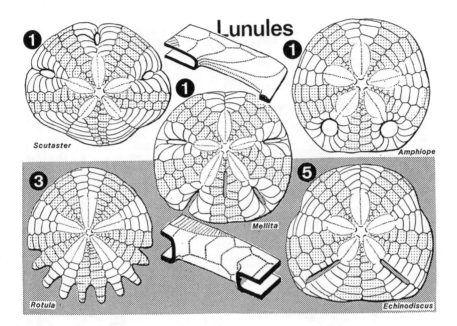

Figure 6. Lunular notches and perforations serve the double purpose of shortcuts for food transport and lift reduction in exposed emergency situations. Nevertheless, this is not an obligatory element of the sand-dollar syndrome. Therefore, it was introduced late in some clades and not at all in others. The independent introduction of lunules in at least six instances of sand-dollar evolution (fig. 8) is emphasized by the choice between two alternative fabricational modes. In mellitids, cross-linkage of upper and lower plates in the interambulacral anal lunule versus festooned arrangement in the interambulacral lunules suggests independent introduction of the two kinds. Independent origins in other examples are derived from chorological and chronological evidence (fig. 8).

**Lunular Notches and Lunules** (fig. 6). Given that sand dollars feed by gravitational sieving on the upper surface, but that they have to live with a ventral position of the mouth, improvement of the food transport must be at a premium. Banana roads have been one step in this direction. Lunular shortcuts are another – all the more since they may also serve additional functions, such as lift-reduction in exposed animals (Telford 1984, 1988), pillar function, or sanitation. Such bizarre perforations would have been impossible to make in a conveyor-belt test. In sand dollars, however, their fabrication posed no major problems, particularly since the sharp margin could be easily held back

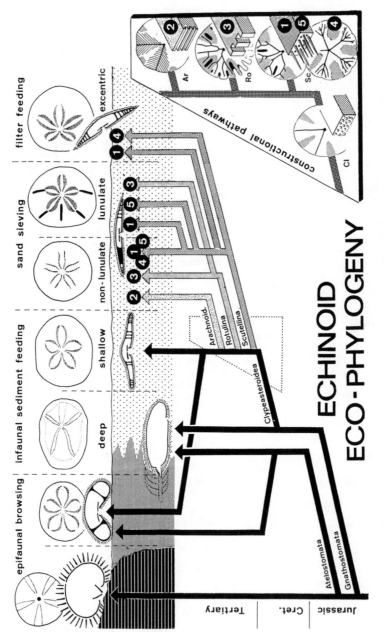

Figure 7. Constraints of the echinoid bauplan and the possibilities of its transformation tend to structure the effective environment into a finite number of ecological zones that can be reached only along certain adaptational pathways. In this evolutionary "road system" the zone of true sand dollars (stippled arrows) forms an extreme, but attractive dead end area. It can be subdivided into subzones that were entered only by part of the recognized clades. Modified from Seilacher 1979.

during growth. The marginal notches thus produced would already have served the functions listed above and could be improved further by marginal closure into true lunules.

In spite of their obvious functionality, lunules are not a necessary element of the sand-dollar syndrome. Lunular notches or perforations appear never to have evolved in the Arachnoidida, nor in the Echinarachniida. *Dendraster* also thrives without such structures, although with a slightly modified mode of feeding. Its eccentricity with respect to the food grooves and the apical system are paralleled in the fossil genus *Scutellaster* (fig. 5), which has an echinarachniid mode of nonlunular food-groove branching. Since lunular structures made their appearance rather late, when sand dollars were already fairly diversified, we may expect a high degree of polyphyly, whose recognition is facilitated by the following fabricational alternatives: (1) position of lunules within ambulacral and/or interambulacral areas, and (2) formation of lunular walls either by direct cross linkage between oral and aboral plates or by "festooning" (fig. 6).

These and other criteria suggest that lunular structures are secondary elements of the sand-dollar syndrome and have independently originated in at least six lineages.

**Chorophylogeny** (figs. 7, 8). The structuralist view of organisms is mainly concerned with developmental systems and with the licenses that constrain their evolutionary transformations. It leads to predictions about the sequence in which changes are likely to have occurred (fig. 7), but not about their actual coordinates in time and space. For this information we rely on biostratigraphic and biogeographic evidence.

The present distribution of sand dollars is well documented and shows a clear pattern. The Scutellina (in the traditional classification) are spread around the Americas and the Indo-Pacific coasts of Asia and East Africa. The Rotulida are restricted to the tropical west coast of Africa. Arachnoididae, in contrast, have their center in Australia and extend into neighboring

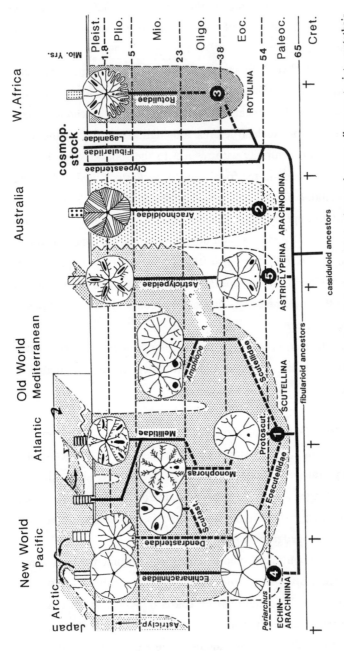

Figure 8. Clypeasteroid choro-phylogeny. The known distribution of clypeasteroids in time and space allows us to interpret their evolutionary history in terms of dispersal and vicariance in an ever changing paleogeographic setting. While the province of the West African Rotulina remained confined by cold upwelling zones, migratory routes for other clades were closed by the widening of the Atlantic and the emergence of the Levantine and Panamanian land bridges. Others opened by collision of Australian and Southeast Asian plates and by the warming of polar seas (arrows). Note that the independence of the Echinarachnina is derived from their food groove pattern and of the Astriclypeina from the precocious appearance of lunulate *Echinodiscus* in Taiwan – unless this island is considered as an exotic terrane of American origin. The family tree also reveals that the explosive evolution of clypeasteroids in the Early Tertiary was favored by the availability of niches after the end-Cretaceous extinction and by the easier dispersibility of the less stenotopic stem groups.

Southeast Asia.

Within the Scutellina, the American realm has the highest diversity at the family level, with the nonlunulate Dendrasteridae and Echinarachniidae becoming cold-adapted and disjunctly spreading to Southeast Asia. The latter group also shows an interesting disjunction at the species level, with *Echinarachnius parma* occurring in the temperate zones both on the West and East coasts of North America. The lunulate Astriclypeidae appear as a Southeast Asian group now extending along the African east coast and into Australia.

From the present picture alone we could postulate centers of dispersal in the Caribbean, West Africa, and Australia, but the spacial derivation of the Astriclypeidae remains unclear.

The paleontological record, although favored by a relatively high fossilization potential of sand-dollar tests, is more scanty. It basically confirms the view of the three spreading centers, but it also adds a flourishing history in the Mediterranean part of the Tethys, where sand dollars had become extinct by the end of the Miocene (Messinian Event).

Figure 8 tries to interpret this picture in terms of available migration routes in a world changing by the broadening of the Atlantic, the emergence of land bridges (Panama, Levante), the collision of formerly separated provinces (Wallace Line), cold upwelling barriers (West African Coast) and changing climates. Although there may be arguments about details (Smith 1984), this picture appears acceptable in a broad sense, illustrating the interaction of dispersal and vicariance in the diversification of a group on a global scale.

There is, however, one record that – if correctly dated – smashes our scenario. It is an astriclypeid sand dollar (*Echinodiscus tiliensis*, Wang 1984) from the Eocene of Taiwan, which already has all the characteristics of an astroclypeid, including fully developed lunules. The occurrence of *Astriclypeus* in the Oligocene of the same area (Wang 1983) increases the credibility of this report. Both records would be too early for a lunulate species to occur at any place and even more so in an area outside the assumed centers of dispersal!

True, the presumed ages of these occurrences must be corroborated by field evidence and independent biostratigraphic data. But if they are confirmed, we must accept a fifth independent origin of the sand-dollar syndrome and a fourth dispersal center in Southeast Asia (fig. 8).

On the other hand, this example reminds us that our cladistic models may always be challenged by a single uncomfortable fossil and that we still have to check diagnostic criteria in all members of the different groups, no matter how homogeneous they might appear.

## Conclusions

The Clypeasteroida are irregular echinoids that share the synapomorphic feature of a rigid but flattened test supported by internal pillars. In terms of *evolutionary ecology* this construction allowed several groups to enter the novel niche of the sand dollars. They sieve fine food particles from a thin veneer of sand on top of the animal and transport them to the mouth, which maintains its traditional position in the center of the lower surface. This innovation implies a syndrome of adaptive features in a functionally defined hierarchy: (1) excessive flattening and smoothing of the test by sharpening the ambitus; (2) division of labor between locomotory spines on the lower coronal surface and sieve spines on the upper; (3) branching of food grooves toward a banana-road paradigm; (4) interlocking of plates as a fortuitous license for extra test strengthening, and (5) introduction of lunular notches and windows as shortcuts for the food transport, with the side effect of hydrodynamic lift reduction in emergency situations.

The view of *constructional morphology* reveals not only constraints in the realization of these adaptive goals, but also alternative fabricational solutions, whose equivalence and irrevocability make them particularly suitable for distinguishing polyphyletic clades. By such criteria the Echinarachniina are recognized as an independent clade of sand dollars in addition to the three already established.

All these cladistic conclusions could have been reached by the study of modern forms alone. The independence of the Astriclypeina as an additional fifth clade, however, is derived from *biohistorical* data. The fossil record also tells us that the sand-dollar innovation took place with an unusual evolutionary tempo. All five groups originated within (or shortly after) the Eocene, only 10-20 million years after the first appearance of less specialized clypeasteroids. The record also suggests that this explosive radiation was not only the outcome of a serendipitous preadaptation, but also of an extrinsic, biohistoric coincidence: the rapid cosmopolitan dispersal of a less specialized stem group into the vacuum left by the end-Cretaceous extinction. Only on this basis could the spectacular sand-dollar syndrome multiply and evolve in different provinces. After this radiation, further optimization proceeded more slowly and at different speeds in the five groups. Two of them never evolved lunules and only two genera have managed to enter still another ecologic zone, namely, that of epibenthic suspension feeders.

Their specialized mode of life confines all sand dollars to shallow marine environments. This ecologic constraint has slowed and channeled the subsequent dispersal in all five groups. It also excluded them from the *offshore escape* under the pressures of biological competition and physical perturbation. Therefore, sand dollars are doomed to eventual extinction, as happened in Europe during the Messinian event.

In a still broader biohistoric context the uniqueness of the sand-dollar syndrome is not just a mere coincidence, because it was possible only after (1) the test had become a rigid capsule in late Triassic euechinoid ancestors and (2) the innovative syndrome of the infaunal Irregularia had been entered in the Lower Jurassic, implying deviation from radial symmetry, adaptation of the spines for a burrowing function, and a stop of ontogenetic plate motion around the ambitus.

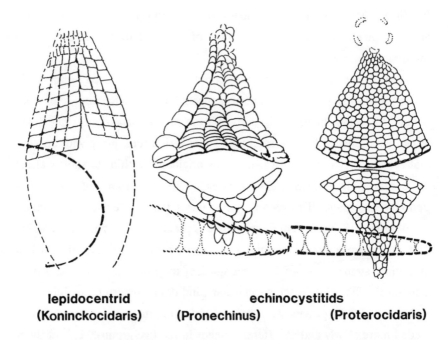

| lepidocentrid | echinocystitids | |
|---|---|---|
| **(Koninckocidaris)** | **(Pronechinus)** | **(Proterocidaris)** |

Figure 9. Paleozoic sand dollars? While the construction of rigid disc-shaped tests required a long series of preadaptations, another window to the sand-dollar niche may have been open much earlier in echinoid evolution. The plated but hydrostatically supported tests of Paleozoic echinozoans could transform into disc shapes, if oral and aboral surfaces were held together by nonmineralized tensional tetherings (dotted lines). In the absence of preserved spines, however, it remains uncertain whether such forms actually lived like modern sand dollars. Modified from *Treatise on Invertebrate Paleontology.*

There might have been, however, another window into the sand-dollar niche at a much earlier point in echinoid evolution. Similarly, disc-shaped echinoids from the Paleozoic (fig. 9) could represent such an evolutionary analogy. At that stage, plates were still flexibly connected, so that the test had to be continuously stabilized by hydrostatic pressure. Also the introduction of new plates was not yet confined to the apical pole. In such flexible capsules tensional internal tetherings must have played a similar role as pillars do in the rigid version. But whether the Paleozoic pancake echinoids were actually infaunal and whether they have used this earlier window to a new ecologic zone can only be verified by other elements of the sand-dollar syndrome, such as sieve spines, food-groove systems, and perhaps even lunules.

In general, our examples show that evolutionary innovations are thresholds that have speeded up evolutions in many instances. But since they are usually part of more complex transformational systems, their key elements may be hard to identify. Also, their ultimate success depends, like everything in life, on the vicissitudes of the historical circumstances. It is probably a mere coincidence that the sand-dollar innovation has been more successful than the float of *Scyphocrinites*!

## Acknowledgments

This is Number 204 of the series *Konstruktionsmorphologie*. Various versions were typed at Yale and Tübingen Universities. I appreciate discussions with Chia-Ching Wang from Yale about Taiwanese sand dollars.

## References

Dafni, J. 1986. A biomechanical model for the morphogenesis of regular echinoid tests. *Paleobiology* 12: 143-60.

Durham, J. W. 1966. Clypeasteroids, pp. 450-91. In *Treatise on Invertebrate Paleontology*. Part U, Echinodermata 3 (2), ed. R. C. Moore. Boulder, CO, and Lawrence, KS: Geological Society of America.

Ghiold, J. 1979. Spine morphology and its significance in feeding and burrowing in the dollar *Mellita quinquiesper=forata* (Echinodermata: Echinoidea). *Bulletin of Marine Science* 29: 481-90.

Ghiold, J. 1984. Adaptive shifts in clypeasteroid evolution – feeding strategies in the soft-bottom realm. *Neues Jahrbuch für Geologische und Paläontologische Abhandlungen* 169: 41-73.

Haude, R. 1972. Bau und Funktion der *Scyphocrinites*-Lobolithen. *Lethaia* 5: 95-125.

Kier, P. M. 1982. Rapid evolution in echinoids. *Paleontology* 25: 1-9.

Seilacher, A. 1979. Constructional morphology of sand dollars. *Paleobiology* 5: 191-221.

Seilacher, A. 1988. Why are nautiloid and ammonite sutures so different? *Neues Jahrbuch für Geologische und Paläontologische Abhandlungen* 177: 41-69.

Seilacher, A., H. Drozdzewski, and R. Haude. 1968. Form and function of the stem in a pseudoplanktonic crinoid (*Seirocrinus*). *Paleontology* 11: 275-82.

Smith, A. 1984. *Echinoid Paleobiology*. London: Allen & Unwin.

Telford, M. 1984. An experimental analysis of lunule function in sand dollar (*Mellita quinquiesperforata*. *Marine Biology* 76: 125-34.

Telford, M. 1988. Ontogenetic regulatory mechanisms and evolution of mellitid lunules (Echinoidea, Clypeasteroida). *Paleobiology* 14: 52-63.

Wang, C. C. 1983. A new species of *Astriclypeus* from the Wuchihshan Formation near Chilung, Taiwan. *Bulletin of the Central Geological Survey* 2: 113-20.

Wang, C. C. 1984. Fossil *Echinodiscus* from Taiwan. *Bulletin of the Central Geological Survey* 3: 107-15.

# The Ecology of Evolutionary Innovation:
# The Fossil Record

*David Jablonski and David J. Bottjer*

Innovation is the mainspring of macroevolution. Other mechanisms come into play once a novelty has arisen, including natural selection and the other sorting processes, at a variety of levels, that drive the waxing and waning of clades and the deployment of long-term trends within clades. However, it is the evolutionary novelty – in one guise or another – that provides the raw material on which those forces operate. The most dramatic kinds of evolutionary novelty, major innovations, are among the least-understood components of the evolutionary process. A number of new – or renascent – approaches offer fresh opportunities for the analysis of the origin of novelties in general and innovations in particular. One, of course, comes from developmental biology (e.g., Bonner 1982; Raff and Kaufman 1983; Goodwin et al. 1983; Raff and Raff 1986; Raff et al., this volume; Müller, this volume). Another comes from paleontology, in which the fossil record is viewed as a document of natural experiments, permitting, within the confines of a historical science, empirical testing of hypotheses on the controls of patterns in the origin of evolutionary novelties.

In this chapter, we discuss paleontological patterns in the origin of innovation that may illuminate process; some authors have suggested that these patterns are regulated or constrained ecologically. We will discuss some large-scale temporal and environmental patterns, and will mention geographic patterns, which we expect to be explored fruitfully in the next few years. A number of workers have shown that major innovations in the fossil record are not evenly distributed in time and space; furthermore, we have

found markedly discordant patterns at different levels in the taxonomic hierarchy, so that a simple extrapolation of rates and patterns at low levels cannot account for the origins of higher taxa.


### Evolutionary Innovation


Evolutionary innovation is to some extent in the eye of the beholder, and
terminology is often confused and contradictory. We prefer a broad definition along the lines of Runnegar (1987:41), that innovation involves "the
crossing of a functional threshold." Thus, a small change in function is still
an innovation, albeit a minor one. No reference is made to the long-term
consequences of this change; by our usage many innovations fail to persist,
let alone trigger diversification (e.g., Jablonski 1986). We will use the term
"novelty" in an even more neutral sense, to denote any derived trait (i.e., apo-
morphy in Hennigian terms), without regard to function or evolutionary consequences. Mayr (1960) frames a similar definition for novelty, but then
"tentatively" restricts the term to evolutionary changes that permit new functions (thereby becoming equivalent to our definition of innovation).

　　If the term innovation must be linked to diversification, then a modifier is called for, and "key innovation" is the logical choice. This term has
long denoted a novelty that permits the exploitation of new habitats or other
resources to the point of triggering an evolutionary radiation (Miller 1949;
Liem 1973); Simpson (1953) uses "key mutation" in a similiar sense, although
this seems to imply a particular kind of genetic basis, but Van Valen's
(1971:422) "key character" is closer to our use of "innovation." This key
innovation concept requires that arguments for the relationship between a
new feature and ensuing diversification go beyond correlation to causation,
a daunting task that requires explicit testing of rival hypotheses (Liem 1973,

1980, this volume; Lauder 1981; Fisher 1985; Levinton 1988:304-9; and see Raikow 1986; Vermeij 1988; and Fitzpatrick 1988 for an interesting exchange). To give a paleontological example, Skelton (1985) tests, against a series of alternatives, the proposition that changes in the ligament growth of the rudist bivalves were the key innovation that set the stage for the great diversification of the group in the Early Cretaceous, by permitting new growth forms and thus new life habits.

This definition of key innovation leaves the thorny problem of diversifications that may be by-products, or in Vrba's term (Vrba 1983; Vrba and Gould 1986) macroevolutionary effects, of novelties rather than the direct results of breakthroughs to vacant niches or to adaptive superiority. Thus, the timing will be right but the functional link will be only indirect. For example, Taylor (1988) has attributed the explosive diversification of the cheilostome bryozoans, some 30 million years (Ma) after their origination, to the appearance of larval brooding, which by reducing dispersal capabilities could greatly increase speciation rates. Similarly, Doyle and Donoghue (1986) suggested that the angiosperm diversification was triggered, at least in part, by closure of the carpel, which would have produced changes in dispersal mechanisms that also could have boosted speciation rates.

As Cracraft (this volume) emphasizes, a critical approach to so-called key innovations is needed; a number of traits long considered under this rubric have proven to be present in far less successful sister taxa as well. Thus, the sister group of the angiosperms had virtually all of the traits held at one time or another to be the key to angiosperm success (Crane 1985; Doyle and Donoghue 1986). Closer to home, functional analysis of the hand of robust australopithecines suggests that this extinct hominid lineage was also capable of precision grasping and tool use, so that "acquisition of tool behavior does not account for the emergence and early success of early *Homo*" (Susman 1988).

### Temporal Patterns

When examining large-scale evolutionary patterns over geologic time and across diverse taxa, many workers have used the first appearance of high-ranking taxa (orders, classes, phyla) as proxies for the origins of major innovations. This approach has its drawbacks, of course: the taxa are not always rigorously delimited phylogenetically, and there is no guarantee that higher taxa of equivalent rank will be comparable across phyla (see Van Valen 1973 for a thoughtful treatment). Nevertheless, this approach has been extremely productive, particularly in analyses of marine invertebrates (e.g., Valentine 1969, 1973, 1980, 1989; Bambach 1983, 1985) – if only because orders of metazoan invertebrates of roughly similar organizational complexity are more equivalent than are, say, one order each of angiosperms, bivalves, and birds. Thus, we disagree with Smith (1984:454) that "nominal categorical rank is unimportant and is best based on diversity or historical precedence." While recognizing the difficulties of classifications incorporating both cladistic relationships (which are, of course, prerequisite to all sound classifications) and some measure of morphologic distinctness, we believe that there are – or can be – functional morphologic underpinnings to the ranking of higher taxa. Bambach (1985) has provided a valuable survey of just this aspect of marine invertebrate higher taxa through the Phanerozoic, illustrating the functional basis of the major groups.

The most striking temporal pattern of evolutionary innovation – or the origin of higher taxa – among skeletonized marine invertebrates (which have the most complete and thus evolutionarily most informative fossil record) is their concentration in the early Paleozoic (fig. 1). This episode of origination at the highest taxonomic levels, which introduced all 11 of the skeletonized marine invertebrate phyla, 54 of the 56 recognized classes, and 152 of the ca. 235 recognized orders (Erwin et al. 1987), along with a number of bizarre forms of uncertain taxonomic placement or rank (Valentine 1977; Brasier 1979; Whittington 1980; Conway Morris 1985; Hoffman and Nitecki 1986),

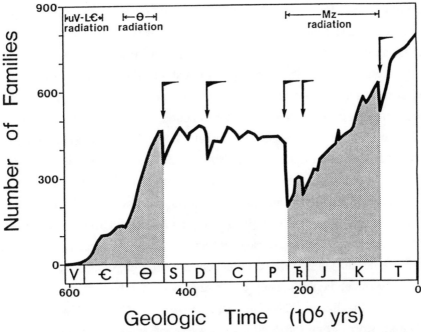

Figure 1. Diversity history of skeletonized marine invertebrate families. Shaded regions represent times of major diversification, with bracketed intervals at top indicating diversification pulses; uV = upper Vendian (= latest Precambrian); LC = Lower Cambrian; Mz = Mesozoic. Arrows indicate the five major mass extinctions. From Erwin et al. (1987), used by permission of the authors and the Society for the Study of Evolution.

has often been called the Cambrian explosion. Although phylum and class-level origination is indeed concentrated in the first 50 Ma of the Phanerozoic, Sepkoski (1979, 1981; see also Brasier 1979, 1982; Runnegar 1982; Sepkoski and Sheehan 1983) has shown that the diversification was spread through the Cambrian and Ordovician in two discrete phases (or may have been a single prolonged episode; see Smith 1988). As Raup (1983) points out, the early appearance of more inclusive taxa is an inevitable consequence of the geometry of phylogenetic trees, but the early appearance of such a broad array of disparate morphologies, as seen in the Cambro-Ordovician diversifications (Valentine 1973, 1977; Whittington 1980; Sprinkle 1983; Valentine and Erwin 1987; Campbell and Marshall 1987) is not.

Valentine (1986), Bengtson (1986), Valentine and Erwin (1987),

Campbell and Marshall (1987), and Erwin et al. (1987) review current hypotheses on the nature and mechanisms of the early Paleozoic radiations. These authors reject the once-popular view that the apparent magnitude and abruptness of the radiations can be attributed to a skeletonization event, in which the simultaneous acquisition of readily-preserved hard parts caused many ancient, long-diverged higher taxa to enter the fossil record all at once (see, for example, Lowenstam and Margulis 1980). Instead, the rapid radiation and elaboration of taxa having body plans that cannot function without durable skeletons (e.g., brachiopods, many echinoderm classes), the coincident increase in bioturbation and diversification of traces produced by soft-bodied taxa (Seilacher 1956; Crimes 1974; Brasier 1979; Droser and Bottjer 1988), the coincident diversification of protists and metazoans with agglutinated rather than biomineralized skeletons (Signor 1988; Lipps 1990) and the apparent radiation of relatively large and complex soft-bodied organisms recorded in the Burgess Shale and similar deposits (Conway Morris 1985; Conway Morris et al. 1987), all suggest a true evolutionary explosion.

The two most attractive hypotheses for the great concentration of evolutionary innovation in the early Paleozoic, potentially rivals but possibly complementary, are ecological and genomic, respectively. The ecological hypothesis (Valentine 1973, 1980; Erwin et al. 1987) holds that the low-diversity marine faunas of the early Paleozoic afforded an ecological setting of greater opportunity for the establishment of divergent morphologies than did later times (Brasier (1979, 1982) argues that the initial radiation was abetted by a global rise in sea level, which would have increased shallow-marine habitat area). Relatively few modes of life were represented at the beginning of the Cambrian (Valentine 1973; Ausich and Bottjer 1982; Bambach 1983, 1985; Bottjer and Ausich 1986), and, as Bambach documents, the Cambro-Ordovician radiation saw not only higher taxa originating within established life-habit categories, but the initial invasion of new modes of life. Diversification occurred not only by increases in within-community diversity, but by differentiation among communities (Sepkoski 1988). Valentine (1973, 1980) hypothesized that as lineages became established and diversified in different ecological roles (adaptive zones sensu Simpson 1953, Van Valen 1971), they tended to preempt later innovations that led other groups to similar or overlapping modes of life, thus increasingly damping the successful establishment of higher taxa, particularly by large or rapid evolutionary steps.

One notable aspect of the ecological hypothesis is the emphasis on preemptive competitive exclusion, in which incumbency evidently permits one clade to suppress or exclude one or more latecomers, as an important force in clade-level interactions. The fossil record does provide more evidence for this mode of interaction than for what Hallam calls displacive competition (e.g., Benton 1983, 1987; Van Valen 1985; Hallam 1987, Gilinsky and Bambach 1987; Miller and Sepkoski 1988; Levinton 1988:463-69); in addition, the diversity models of Valentine (1980; Valentine and Walker 1986; Walker and Valentine 1984) based on this assumption provide a good simulation of real diversity patterns. Further exploration of this aspect of clade interaction is certain to be fruitful. This is not to say that displacive competition among higher taxa never occurs; many authors have interpreted the winnowing of higher taxa after the early Paleozoic burst (e.g., the reduction of echinoderm classes from an Ordovician peak of about fifteen to the present-day five or six) in terms of experimentation followed by exclusion of the less efficient designs. Paul (1977, 1979) has been the strongest advocate of this view, but Campbell and Marshall (1987) contend that it is not yet proven that the small classes lost in the early and middle Paleozoic were inferior in design to those that persisted; after all, there are many reasons for small clades to become extinct (see Sepkoski 1981; Raup 1983; Strathmann and Slatkin 1983).

A second important, indeed central, implication of the ecological hypothesis is that innovation has occurred at roughly constant rates and magnitudes throughout the Phanerozoic (also explicitly stated by Wright 1982a, 1982b, and references therein). Thus, the long-term taxonomic patterns reflect the restriction of ecological opportunities as the biosphere becomes more densely occupied. This is where the genomic hypothesis differs.

The genomic hypothesis (Valentine 1986; Valentine and Erwin 1987) holds that genomes were less highly canalized and had fewer epistatic (among-gene) interactions in early Paleozoic metazoans than in later forms, so that major alterations in development were more readily accomplished. Ecological opportunities were important, but in drawing upon a looser genomic organization the diversification also entailed "mechanisms of genome change that do not operate at the same intensity or with the same results today" (Valentine and Erwin 1987:100). Thus, there is a real temporal bias in the generation of major innovations, dictated by evolution in genome organization.

Unfortunately, these two hypotheses yield few unique predictions testable in the fossil record. The ecological hypothesis might predict that any interval of low standing diversity would provide opportunities for the establishment and elaboration of major innovations and the origin of new higher taxa. Because the end-Permian mass extinction probably removed over 95% of the extant species (see Sepkoski 1989 and references therein), the Mesozoic rebound occurred in an impoverished setting comparable in terms of species richness to that of the Early Cambrian. However, as Erwin et al. (1987) demonstrate, the Mesozoic diversification did not foster a burst of innovation comparable to that of the early Paleozoic: familial diversity attained unprecedented levels, but no new phyla or classes, and significantly fewer orders appear in the post-Paleozoic interval. The differences between the two radiations might be taken as support for the genomic hypothesis, with more strictly bound Mesozoic genomes capable of less innovation, but Erwin et al. (1987) argue otherwise: early Mesozoic "adaptive space" did not present the same sort of ecological vacancy as the early Paleozoic. Although greatly impoverished taxonomically, the survivors of the end-Permian extinction were ecologically diverse; Erwin et al. (1987) calculate that 15 of the 20 marine adaptive types delineated by Bambach (1983) were present in the Triassic, actually a slight gain over the Paleozoic tally. The Mesozoic rebound was seeded by survivors scattered over a wide range of life habits and body plans (see Ausich and Bottjer 1985; Erwin et al. 1987). Thus the differences between the two great Phanerozoic radiations can be explained ecologically.

Originations of taxa below the class level do respond to mass extinctions, also in a fashion consistent with the ecological hypothesis (see review by Jablonski 1986). Families undergo a dramatic burst of origination after each mass extinction (Sepkoski 1984; see also Sheehan 1982 on the brachiopod evolutionary response to the end-Ordovician extinction, and Miller and Sepkoski 1988 on the dynamics of origination in bivalve mollusks), and orders arise significantly more frequently following mass extinctions than expected by chance alone (Erwin et al. 1987).

None of these observations falsify the genomic hypothesis, however. The phylum- and class-level burst of innovation in the early Paleozoic is a unique event in metazoan history and may yet demand a unique explanation. Again, the problem is the lack of paleontologically testable predictions. The marked phenotypic variability of many Cambrian taxa compared to younger relatives (e.g., Bergström and Levi-Setti 1978; McNamara 1983, 1986a; Runnegar and Bentley 1983; Conway Morris and Fritz 1984; Andres 1988) might reflect less canalized genomes, but they could also reflect relaxed normalizing selection in a relatively empty adaptive landscape. One unexplored possibility is a comparison of fluctuating asymmetry between Cambrian and younger populations. Such small, random deviations from bilateral symmetry reflect the strength of developmental homeostatic mechanisms (Van Valen 1962; Palmer 1986; Palmer and Strobeck 1986), and a progressive decrease in such deviations during the history of clades that originated in the early Paleozoic diversification (but not in later times) might provide indirect clues to genome evolution.

Similar temporal patterns in diversification occur at lower taxonomic levels, with increased origination within orders and families during low-diversity intervals, be they the initial early Paleozoic vacancy or post-extinction impoverishment. In the early Paleozoic, within-clade diversity histories tend to be "bottom-heavy," with families tending to arise early in the history of their orders and genera tending to arise early in the history of their families. In contrast, post-Ordovician clades tend not to show such temporal biases (Gould et al. 1977, 1987; but see Kitchell and MacLeod 1988, who maintain the differences are not significant). Van Valen and Maiorana (1985; Van Valen 1985; see also Kitchell and Pena 1984) found significant bursts of familial origination within classes in both the early Paleozoic and early Mesozoic intervals for marine organisms, and generic and familial origination were exceptionally high for the mammals in the early Cenozoic following the opening of adaptive space with the demise of the dinosaurs (see Van Valen and Maiorana 1985; Gould et al. 1987).

Gilinsky and Bambach (1987) go one step further and contend that most orders exhibit an initial burst of familial origination, regardless of time of origin. However, it is not clear that this observation is independent of the association with the various mass extinctions noted above; after all, most orders first appear during the early Paleozoic or in post-extinction rebounds. In alternatives analogous to the arguments presented above, Gilinsky and Bambach (1987) at least entertain the possibility of genome rejuvenation at lower taxonomic levels (i.e., lessening of canalization with the origin of a major innovation or invasion of an adaptive zone); alternatively, they suggest that a formerly advantageous innovation loses its edge as the surrounding biotic and physical environment changes, leading to a decrease in within-clade origination. However, it seems at present that the simplest explanation for these diverse data, across the taxonomic hierarchy, is that ecological opportunities imposed by major perturbations play a key role in the timing and survival of evolutionary innovation (as Wright 1949, among others, suggested long ago).

## Environmental Patterns

Time-independent ecological patterns in major innovation can also be recognized in the fossil record. One of the most intriguing is the preferential origin of higher taxa of benthic marine invertebrates in onshore settings. Expectations might have put originations in more stable environments further offshore or, if successful innovation is largely a matter of the chance combination of novelty and opportunity, in a bathymetrically random distribution. Instead, a significant onshore bias is found in the first appearances of broad faunal associations, both in Paleozoic and in post-Paleozoic marine benthic organisms. Here, we further document these onshore-offshore patterns from a more phylogenetic viewpoint, with an analysis of post-Paleozoic orders of benthic echinoderms.

Sepkoski and Sheehan (1983; Jablonski et al. 1983; Sepkoski and Miller 1985; Miller 1988; Conway Morris 1989; see also Brasier 1982:117-18) document onshore origin and subsequent spread across the continental shelf of major faunal associations during the early Paleozoic. The relationship of these Cambro-Ordovician faunal patterns to the beginnings of the Cambrian explosion are unclear. Mount and Signor (1985) argued that the initial appearance of skeletonized organisms in the earliest Cambrian of California showed no environmental bias; this may well be true, but as Bottjer and Jablonski (1988) point out, in the scheme used here the Cambrian influx of skeletonized forms occurs in settings classified primarily as onshore.

**Post-Paleozoic Orders.** In order to disentangle community versus clade-level processes, and to gather a data set sufficiently large and detailed (temporally and environmentally) for testing rival hypotheses, we have begun to assemble environmental histories of post-Paleozoic orders. First occurrences and later fates of these groups in the fossil record are analyzed within five broad subdivisions of an environmental transect from the intertidal zone to the slope and deep basin. These environmental subdivisions, recognized on the basis of physical sedimentary evidence reflecting environmental energy levels and other depth-related factors, are reviewed by Bottjer and Jablonski (1988; see also Jablonski and Bottjer 1990; Sepkoski [1987, 1988] uses a similar scheme with one additional subdivision). Negative data – absences from a particular point in the time-environment matrix – are evaluated using taphonomic control taxa (Bottjer and Jablonski 1988), which have transport and preservation characteristics similar to the group of interest. Thus, absences of tellinacean bivalves are recorded only if other taxa of relatively small and infaunal bivalves, with shells composed of dissolution-prone aragonite, are present at the relevant locality; cheilostome bryozoan absences are recorded only if cyclostome bryozoans are present; and isocrinid crinoids only if other articulated echinoderms (ophiuroids, asteroids, other crinoids) are present.

Applying these techniques, we find that 15 out of 22 orders of crinoids and echinoids with a post-Paleozoic origin first appear in onshore (= nearshore plus inner shelf) settings. (Ophiuroids, asteroids, and holothuroids have too incomplete a fossil record to be useful in this analysis; see Jablonski and Bottjer 1988.) Among the crinoids, some orders, like the Isocrinida, Millericrinida, and Cyrtocrinida, begin onshore, expand offshore, but then retreat into exclusively offshore habitats, so that today they are scarce at depths shallower than 100 m (Rasmussen 1978; Roux 1987; Jablonski and Bottjer 1988, 1990). In contrast, the order Comatulida, the unattached feather stars, began onshore and expanded outwards, but have never suffered onshore extirpation. These are the great successes of the post-Paleozoic crinoids, occurring relatively commonly from the intertidal zone to abyssal depths (Meyer and Macurda 1977; Macurda and Meyer 1983). The bourgueticrinids present a taxonomic problem. Some authors (including Rasmussen 1978) accord them full ordinal rank, while others consider them of lower rank within the Isocrinida or the Millericrinida (reviewed by Jablonski and Bottjer 1990). Because of this uncertainty, we perform our analyses with and without the bourgueticrinids as a distinct order.

Echinoid orders exhibit a pattern qualitatively similar to that of the crinoids (Jablonski and Bottjer 1988, 1990). The larger sample size permits two further observations: the echinoids appear very much a part of the early Mesozoic rebound from the Permo-Triassic mass extinction, and the onshore bias in origination is a statistical trend, not an absolute law. For example, the disarmingly named disasteroids may follow a contrary environmental history, first appearing in continental slope deposits in the late Jurassic of southeastern France and spreading upwards onto the shelf from there (see Jablonski and Bottjer 1988, 1990). Nevertheless, most echinoid orders begin onshore and expand outwards, including the oldest irregular echinoids (which first appear in oolitic carbonates of the Early Jurassic Sunrise Formation of Nevada) and the clypeasteroids discussed by Seilacher (this volume).

The environmental pattern in the origin of higher taxa is not restricted to the echinoderms. Bottjer and Jablonski (1988) tabulate a diverse assemblage of arthropod, coelenterate, and molluscan taxa (of varying rank) for which at least some anecdotal evidence suggests origination onshore. Even groups so divergent from echinoids as the bryozoan order Cheilostomata, minute colonial invertebrates that did not appear until the latest Jurassic, well after the heyday of the Mesozoic rebound, first occur in onshore strata (Bottjer and Jablonski 1988; Jablonski and Bottjer 1990)(fig. 2).

**A Discordant Pattern at Lower Levels.** The origin of novelties within these higher taxa contrasts significantly with environmental patterns shown by the orders themselves ($p < 0.005$ if bougueticrinids are an order, $p < 0.001$ if bourgueticrinids are isocrinids; chi-square tests detailed by Jablonski and Bottjer 1990). After the initial establishment of the clade, isocrinid genera tend to appear first in offshore settings (see Jablonski and Bottjer 1990)(fig. 3). The pattern is even more striking if the bourgueticrinids are subsumed within the Isocrinida. The few families involved also fail to show an onshore bias: the first two, the Holocrinidae and Isocrinidae, first appear onshore, but the two younger families recognized by Simms (1988a), Cainocrinidae and Isselicrinidae begin in the middle or outer shelf, as do the two bourgueticrinid families. The family data are too few for statistical analysis, but the first occurrences of genera are significantly different in their environmental distribution from the echinoderm orders (fig. 4). We could not have predicted the

# CHEILOSTOMATA

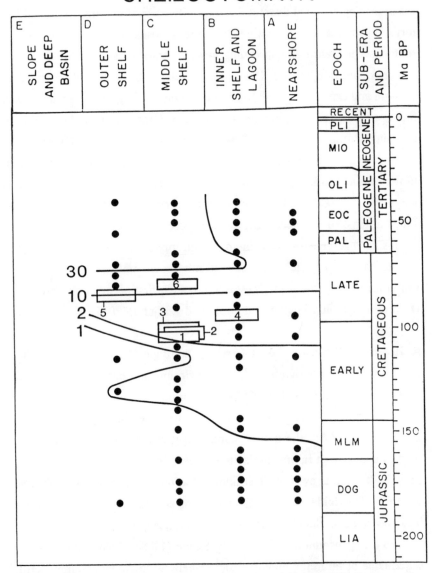

Figure 2. Environmental history of the bryozoan order Cheilostomata in the Euramerican region, with contours showing within-habitat generic richness, and environment of first occurrence of six novelties. Key to novelties: 1, ovicells; 2, erect colony form; 3, cribrimorph organization; 4, articulated colony form; 5, ascophoran organization; 6, lunulitiform colony form. From Jablonski and Bottjer (1990), who define and document novelties. See Bottjer and Jablonski (1988) for documentation of environmental history.

# ISOCRINIDA

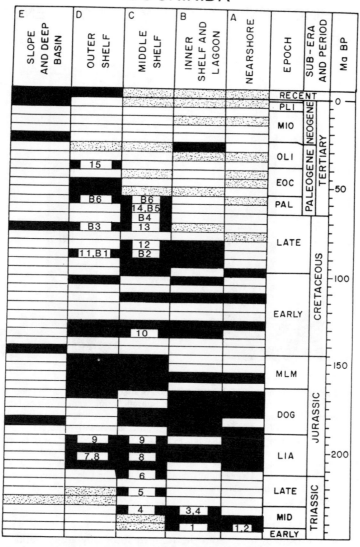

Figure 3. Environmental history of the crinoid order Isocrinida in the Euramerican region, and environments of first occurrence of isocrinid and bourgueticrinid genera. Solid boxes represent presences, stipples represent controlled absences. Key to isocrinid genera: 1, *Holocrinus;* 2, *Moenocrinus;* 3, *Laevigatocrinus;* 4, *Isocrinus;* 5, *Singularicrinus;* 6, *Chladocrinus;* 7, *Balanocrinus;* 8, *Hispidocrinus* Simms (1988b); 9, *Chariocrinus;* 10, *Nielsenicrinus;* 11, *Austinocrinus;* 12, *Isselicrinus;* 13, *Doreckicrinus;* 14, *Cainocrinus;* 15, *Tauriniocrinus.* Key to bourgueticrinid genera: B1, *Bourgueticrinus;* B2, *Monachocrinus;* B3, *Democrinus;* B4, *Dunnicrinus;* B5, *Bathycrinus;* B6, *Conocrinus.* From Jablonski and Bottjer (1990), who document generic occurrences. See Bottjer and Jablonski (1988) for documentation of time-environment history.

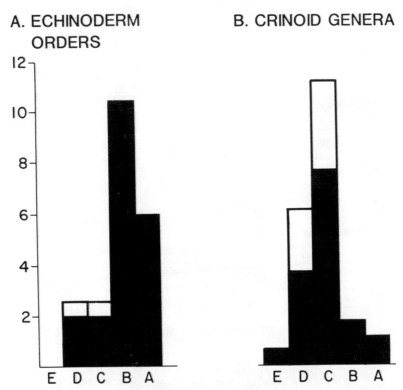

Figure 4. Summary of onshore-offshore distribution of first occurrences of (A) post-Paleozoic crinoid and echinoid orders, and (B) isocrinid and bourgueticrinid genera. Open boxes represent treatment of Bourgueticrinida as a distinct order (in A) or as part of the broader isocrinid clade (in B). The distributions of first occurrences are significantly different (Kolmogorov-Smirnov test, $P$ < 0.05 with bourgueticrinids as a distinct order, $P$ < 0.005 with bourgueticrinids as Isocrinida; chi-square test with environments grouped into onshore and offshore categories, $P$ < 0.005 with bourgeuticrinids as distinct order, $P$ < 0.001 with bourgueticrinids as Isocrinida). Environments A (nearshore) through E (slope and deep basin) as indicated in figures 2 and 3. Taxa whose occurrences are only resolved to two environments are scored as 0.5 in each environment.

pattern at the ordinal level from the pattern shown by genera and families.

The familial and generic taxonomy of the Cheilostomata is presently uncertain, and here we have gone directly to the within-clade innovations discussed by Voigt (1985)(fig. 2): ovicells (where larvae are brooded; see also Taylor 1988), erect growth form (escaping the constraints and perils of an exclusively encrusting growth habit), erect growth form with an organic, flexible stalk (permitting colonization of turbulent waters), lunulitid growth form (a discoidal, free-living and commonly mobile colony shape), cribrimorph structure (with each individual zooid having an arched wall of spines that fuse to create a sieve-like, calcified shield), and acophoran structure (with a solid, calcified frontal shield in the individual zooid, and a flexible sac providing hydrostatic control). Each of these innovations probably evolved more than once within the Cheilostomata (see Jablonski and Bottjer 1988), but, as with the crinoid genera and families, their first occurrences show no onshore bias.

**Causal Mechanisms.** The temporal and environmental patterns of ordinal origination, and their discordance with novelties at lower levels, allow us to rule out several potential explanations for the onshore bias in the first appearance of major innovations (see Bottjer and Jablonski 1988, and Jablonski and Bottjer 1990, for fuller discussion). First, and by no means a trivial matter, the fact that a few orders show a contrary pattern, and that cheilostome and crinoid within-clade novelties show significantly different patterns from the orders, suggests that the predominantly onshore bias in ordinal origination is not an artifact of preservation or other biases of the stratigraphic record. Still awaiting quantification, however, is effect of variations in rock volume available for sampling in each environmental category.

Second, few clades appear to begin simultaneously, even at our coarse 5-Ma resolution, and they tend to move across the shelf at different rates; this argues against community-level processes (Jablonski et al. 1983). Community-level analyses can be deceptive. For example, Bretsky (1969) and McGhee (1981; Sutton and McGhee 1985) found that onshore Paleozoic community types tended to be more stable in composition than offshore community types. But when Sepkoski (1987) decomposed these communities into constituent taxa, the pattern disappeared (see also Thayer 1973), and members of some higher taxa showed increasing generic longevity in an offshore direction.

Third, although many orders begin in the early Mesozoic, not long after the Permo-Triassic debacle (and in an interval punctuated by the late Triassic extinction), mass extinctions apparently did not impose an *environmental* bias. Given the preemptive competition model discussed above, mass extinctions would have to clear onshore environments preferentially in order to provide a setting that would concentrate major innovations; however, there is little evidence for such an environmental bias in severity for the Permian or post-Paleozoic mass extinctions. Some Paleozoic extinctions may have affected onshore taxa more severely, but in these instances offshore stocks have been the source of rediversifications, rather than in-situ innovation (e.g., Stitt 1975, Fortey 1983, Westrop and Ludvigsen 1987 on trilobites; Ausich et al. 1988 on blastoids; and, during normal times, Chauff 1985 on conodonts and - Sando 1980 on corals). However, none of these examples extend to ordinal levels, and counterexamples are readily available (e.g., Anstey 1986 on bryozoans; Frey 1987 on bivalves; Eckert 1988 on crinoids).

Fourth, the discordance between ordinal and within-order originations argues against some of the more straightforward models based on extrapolation from lower levels. It would be logical, for example, to attribute the onshore bias to the diversity of selection pressures that presumably operate there. The greater diversity of selection pressures might be expected to elicit a greater amount of evolutionary divergence onshore than in the more continuous and stable offshore environments. But if this occurred, an onshore concentration of new genera and families would also be expected, as this greater tendency for divergence manifested itself across the hierarchy. Instead, we see discordance across the hierarchy. Within-clade novelties appear to follow constraints imposed by the diversity gradient of the clade: the most species-rich environments tend to give rise to more species and genera (see Jablonski and Bottjer 1990, who also analyze first occurrences of families in tellinacean bivalves).

We are left with two explanations for the onshore-offshore patterns, and these remain to be tested. The first possibility is that major innovations actually arise randomly with respect to environment, but survive long enough to become established, or detected in the fossil record, primarily in onshore environments. The few analyses available for post-Paleozoic invertebrates suggest that onshore species tend to be widespread and physiologically tolerant (Jackson 1974; Jablonski and Bottjer 1990, and references therein), and thereby tend to be extinction-resistant (but see Sepkoski 1987 for contrary examples for Paleozoic genera). Thus, any innovation that happens to appear in one of these extinction-resistant, onshore species will have a greater probability of surviving the perilous, low-diversity phase of diversification than will an innovation of similar adaptive value that appeared among offshore species, which tend on average to be more extinction-prone. This indirect macroevolutionary influence of processes at the organismic or species level might fall under the general heading of effect macroevolution (Vrba 1983; Vrba and Gould 1986).

Alternatively, taking the fossil record at face value, there may be something about onshore environments that evokes major innovation. This is the most exciting possibility, but as we noted at the beginning of our chapter, it is by far the least understood. Jablonski and Bottjer (1983) mentioned, but did not rigorously develop, a population-genetic approach based on Templeton's (1980, 1986) view of founder speciation. However, it is unclear whether the great diversity of taxa involved in onshore innovations possess the appropriate populational traits required for Templeton's transiliences (Jablonski and Bottjer (1990) find that the few available data are encouraging but inadequate for generalization). Returning to the developmental theme of this symposium, perhaps heterochrony (and thus rapid, coordinated steps to innovation) is more common onshore (McKinney 1986). Gould's (1977) ecologic models would require paedomorphosis via progenesis as the primary heterochronic mode onshore (see McNamara 1986b for a useful review of terminology). Although a detailed investigation is needed, Jablonski and Bottjer (1990) suggest that the presence of several non-paedomorphic onshore orders and paedomorphic offshore ones undermines the heterochrony hypothesis.

A multitude of alternative hypotheses remain, most of which are, in principle, testable. For example, perhaps new higher taxa tend to originate from opportunistic species (Margalef 1968, and, for marine benthos, Hermans 1979; and see below), which are most common in disturbed onshore habitats. On the other hand, perhaps we should turn the question around and ask whether offshore environments in some way dampen the origin of higher taxa but do not suppress the production of genera and families. Whatever the outcome of these tests, we can conclude that the fossil record offers us some unexpected ecological patterns in the origins of major innovations that should provide fertile ground for theory and empirical analysis in the future.

**Terrestrial Organisms.** Present-day terrestrial communities are better understood ecologically than are subtidal marine communities, but unfortunately the terrestrial fossil record is probably inadequate for the kinds of time-environment analyses we have presented here for marine taxa. The vertebrate record is so highly skewed towards preservation in a few depositional environments (Behrensmeyer and Hill 1980; Behrensmeyer 1984; Behrensmeyer and Kidwell 1985; Kidwell and Behrensmeyer 1988), that such analyses may never be practical.

The plant record, particularly of pollen, may be more suitable, but until recently little work had been done. The long-standing contention that angiosperms originated in upland environments long before they entered the fossil record (Axelrod 1952, 1961) was based mainly on the apparent derived nature of early angiosperms and the absence of obvious precursors in floodplain deposits, rather than on direct evidence. Recent work on early angiosperms does suggest an intriguing ecological pattern, with angiosperms originating as weedy, fugitive species in disturbed streamside habitats (Doyle 1978, 1984; Tiffney 1984; Retallack and Dilcher 1986; Crane 1987). This is consistent with one of the suggestions we noted above for marine invertebrates. In a pattern (but seemingly not a process) analogous to the onshore-modern, offshore-archaic pattern seen for marine invertebrates, Knoll (1985) suggests that swamps and other wetlands have been refuges for archaic plant taxa throughout the Phanerozoic, rarely the source of innovations relative to uplands and floodplains. Knoll explains this in terms of the relatively limited area of these rather extreme habitats, and the evolutionary commitments of swamp taxa to specializations for this wet, acidic, oxygen-poor, and nutrient-poor environment that would tend to hamper radiation outside this habitat. Time-environment histories of major plant groups in their phylogenetic context may yield general insights into the ecological origins of innovation.

## Geographic Patterns

Paleobiogeographic patterns in the origin of higher taxa have received relatively little attention in recent years. Data are still scarce and unevenly distributed, but ongoing fossil collection and description outside the well-sampled Euramerican region and advances in plate tectonic reconstruction make the geographic analysis of first occurrences in the fossil record an increasingly feasible prospect. We refer not to a new search for centers of origin, as has been emphasized in the past (see Nelson and Platnick 1981 and Briggs 1984 for reviews from opposing standpoints), but instead we hope that new insights into the origin of major innovation can develop from geographic patterns (or the lack thereof) in first-occurrence data. At this early juncture, we can discuss only the most general biogeographic pattern in today's biota, the latitudinal diversity gradient from poles to tropics. The longitudinal diversity gradient, along the equator towards a diversity maximum in the Indo-West Pacific, is almost as impressive (Valentine 1973; Vermeij 1978), but even fewer relevant paleontological data are available.

Because modern latitudinal diversity gradients are evident for genera and families as well as species (Stehli 1968; Campbell and Valentine 1977), it has generally been assumed that taxa at these ranks tend to originate in the tropics. This interpretation was supported by the discovery that mean generic age within major groups (bivalves, corals, benthic and planktic Foraminifera) decreased with latitude (Stehli et al. 1967, 1969, 1972; Stehli and Wells 1971). The logical implication is that large geographic area, high species richness, and – it was assumed – stable environment all combined to foster the production and accumulation of many new species, genera, families, and on up the hierarchy. However, two kinds of data complicate straightforward interpretations.

First, Stehli et al. (1972) argued that averages ages of extant planktonic foraminiferal species showed no latitudinal trend, even though the average age of extant genera becomes younger towards the tropics. Their data, however, do in fact reveal a significant decrease in species age with latitude (J. J. Sepkoski, Jr., pers. com.). Further analysis by Wei and Kennett (1986), on the other hand, failed to find significant differences in speciation, extinction, and turnover rate in a much larger set of living and extinct Neogene Foraminifera (they provide no data at the genus level). Stehli et al. (1972) took their data to indicate that favorable high temperatures in tropical seas permitted innovation ("crossing an adaptive threshold") to occur more readily at low latitudes, whereas reproductive isolation and thus speciation did not depend on crossing such thresholds, so that speciation probability was independent of latitude. In some ways, their explanatory model does not go far beyond restating the observations in ecological terms, but if the basic trends can be further confirmed, they represent an intriguing pattern that deserves much more investigation.

Second, Rosen (1984) found that the tropical Indo-West Pacific minimum in average age did not represent a fauna composed entirely of young genera, but instead comprised an assemblage of old genera combined with a group Middle Miocene or younger in age (demonstrating again the utility of stratigraphic range frequency distributions relative to mean values, which could, of course, be underlain by many different kinds of age distributions). The difference between the Indo-West Pacific and regions at other latitudes and longitudes lay in the relative production or accumulation of these young taxa, rather than in discrete age distributions among regions. Valentine (1984) suggests that this Neogene (i.e., Miocene and younger) diversification may be a general phenomenon for the shallow-sea biota. Recent data on Cenozoic paleotemperatures indicates a late Oligocene-early Miocene interval of significant tropical cooling, with low-latitude Pacific sea-surface temperatures near 19° C, rather than today's 28°-30° (Douglas and Woodruff 1981; papers in Kennett 1985; Wei and Kennett 1986; but see Matthews and Poore 1980 and Poore and Matthews 1985, who attribute isotopic values to fluctuations in polar ice volume rather than temperature variations; and Miller et al. 1987 for a synthesis). At the same time, extinctions occurred in corals (Rosen 1984) and other marine invertebrates (Raup and Sepkoski 1986; Sepkoski 1986), as well as in the tropical planktic Foraminifera (Hoffman and Kitchell 1984; Wei and Kennett 1986) that were the basis of the species vs. genus differences in survivorship. As Valentine (1984) suggests, biogeographic patterns in both standing diversity and origination rates can be attributed to radiations during the past 15 Ma in the wake of a perturbation most strongly felt in the tropics, rather than to a long-standing, stable situation. Whereas the Mediterranean region and the Caribbean were prolific sources of new coral taxa when they were part of the vast tropical Tethyan region, at present they are much smaller and more isolated, and hence no longer favorable settings for diversification, relative to the enormous Indo-West Pacific (see Rosen 1984; also Stanley 1979 on mollusks).

Latitudinal evolutionary patterns above the genus level are even less well known. In principle, each family, order, and class begins with a single species in a single region, but there are great problems with uneven sampling and taxonomy. Most authors seem to agree with Gordon (1974:145) that "observational data and theoretical considerations both show the [tropical] Tethys Sea was the principal center for evolution and dispersal of successful new marine species and higher taxa during Mesozoic time." However, this conclusion – which we agree may well be correct – could in part be due to biases in collection intensity and outcrop availability. A number of marine invertebrate genera and families are now known to first appear in the fossil record at high latitudes and expand towards the equator (Zinsmeister and Feldmann 1984; Crame 1986). Hickey et al. (1983) argued that a number of terrestrial plant and animal taxa exhibit the same pattern, but the stratigraphic correlations on which these conclusion were based have been questioned (Flynn et al. 1984; Spicer et al. 1987). The few marine data for high-latitude origination are more firmly based, but we do not yet know if they represent a general pattern or idiosyncratic histories of relatively few taxa. Further, few data exist above the family level. Taylor et al. (1980) argue that the Order Neogastropoda, and most of its constituent families, originated in extratropical latitudes (see also Sohl 1987), and this prolific order only later achieved its present-day tropical diversity maximum. How many other groups have followed such a route? In Stenseth's (1984) phrase, are the tropics a cradle or a museum?

More data are needed before we can generalize concerning latitudinal patterns in the origins of higher taxa. Once such data are gathered, in a fashion that is robust to sampling and other problems, they should provide a new basis for understanding the ecology of innovation. For example, Stenseth (1984) develops a Red Queen model for preferential high-latitude origination of higher taxa, building on the assumption that the dense packing of species at low latitudes causes high tropical speciation rates to be matched by high extinction rates. From our own preliminary survey of post-Paleozoic echinoderm orders (Jablonski and Bottjer 1988), we are impressed that twelve out of twenty-two are first recorded from deposits of the Tethyan tropical seaway, given the much scantier sampling outside of North America and Europe. These data are not yet sufficiently refined to draw firm conclusions, and ongoing high-latitude collection or low-latitude taxonomic revision might change the results, but we are confident that biogeographic analyses can make a substantial contribution to the study of innovation (see also Jablonski et al. 1985).

## Conclusions

Major evolutionary innovations are not randomly distributed in time and space. Temporally, higher taxa such as orders (which we take, with some trepidation, as indicators of innovation) have first appearances concentrated when global diversity is low, at the onset of the early Paleozoic metazoan radiations and in the wake of mass extinctions; ecological opportunity is the most obvious explanation. Spatially, low-diversity settings are less clearcut sources of higher taxa. Onshore settings preferentially give rise to new orders, and Sheehan (1986) has argued that low diversity here is also the key, with less competition among resource specialists to prevent innovations from gaining a foothold; as we have discussed, a multitude of alternative hypotheses are available. Regarding latitudinal patterns, Stenseth (1984) and Sheehan (1986) also suggest that low-diversity settings might be more conducive to innovation, but the empirical evidence regarding high-latitude innovation is ambiguous and may even favor high-diversity tropical settings as sources of innovation.

In tracing the environmental histories of evolutionary innovations in post-Paleozoic marine invertebrates, we have documented that a significant number of higher taxa, representing a wide range of life habits, originated in shallow-water environments and later expanded outwards across the continental shelf. This common, but by no means universal, onshore bias in ordinal origination is not simply a dampened or extrapolated reflection of the ecological patterns we see at lower levels, where the production of novelties seems to be diversity-dependent rather than environment-dependent. We are not suggesting that orders arise in single jumps by massive macromutations, nor do we claim that the present taxonomy is so objective that ordinal rank is strictly comparable across phyla. But different patterns imply different processes. Evidently some aspect of the onshore environment fosters production, or enhanced survivorship, of morphologies sufficiently distinct or with sufficient long-term evolutionary productivity that many marine invertebrate groups afforded high taxonomic rank are traceable to onshore species. In terms of the ecology of their evolutionary origins, higher taxa seem to have properties all their own.

## Acknowledgments

We are grateful to many paleontologists who have shared their knowledge and insights. These include: on crinoids, M. J. Simms; on echinoids, B. Clavel, A. B. Smith, S. Suter; on bryozoans, A. H. Cheetham, S. Lidgard, M. R. A. Listokin, F. K. McKinney, M. J. Nowicki, P. D. Taylor, E. Voigt; on environmental problems, A. Hallam, S. M. Kidwell, A. Oravecz-Scheffer, D. B. Rowley, A. M. Ziegler. Susan M. Kidwell, J. John Sepkoski, Jr., and James W. Valentine provided helpful comments on the manuscript. We also thank the staffs of the following libraries, which were indispensable resources for this work: Hancock Library, University of Southern California (especially Suzanne Henderson); Geology and Geophysics Library, UCLA (especially Michael Noga); Field Museum of Natural History Library; and the John Crerar Library, University of Chicago (especially Kathleen Zar). We acknowledge the Donors of the Petroleum Research Fund, administered by the American Chemical Society, for support of this research. Additional support was provided by the Visiting Scientist Program of the Field Museum of Natural History (to DB), and by the National Science Foundation (Grants EAR 85-08970 and EAR 85-19941 to DB; EAR 84-177011 and INT 86-2045 to DJ).

# References

Andres, D. 1988. Strukturen, Apparate und Phylogenie primitiver Conodonten. *Palaeontographica* A200: 105-52.

Anstey, R. L. 1986. Bryozoan provinces and patterns of generic evolution and extinction in the Late Ordovician of North America. *Lethaia* 19: 33-51.

Ausich, W. I., and D. J. Bottjer. 1982. Tiering in suspension-feeding communities on soft substrata throughout the Phanerozoic. *Science* 216: 173-74.

Ausich, W. I., and D. J. Bottjer. 1985. Phanerozoic tiering in Phanerozoic communities on soft substrata, pp. 255-74. In *Phanerozoic Diversity Patterns*, ed. J. W. Valentine. Princeton, NJ: Princeton University Press.

Ausich, W. I., D. L. Meyer, and J. A. Waters. 1988. Middle Mississippian blastoid extinction event. *Science* 240: 796-98.

Axelrod, D. I. 1952. A theory of angiosperm evolution. *Evolution* 6: 29-60.

Axelrod, D. I. 1961. How old are the angiosperms? *American Journal of Science* 259: 447-59.

Bambach, R. K. 1983. Ecospace utilization and guilds in marine communities through the Phanerozoic, pp. 719-46. In *Biotic Interactions in Recent and Fossil Benthic Communities,* ed. M. J. S. Tevesz and P. L. McCall. New York: Plenum.

Bambach, R. K. 1985. Classes and adaptive variety: The ecology of the diversification of marine faunas through the Phanerozoic, pp. 191-253. In *Phanerozoic Diversity Patterns,* ed. J. W. Valentine. Princeton, NJ: Princeton University Press.

Behrensmeyer, A. K. 1984. Taphonomy and the fossil record. *American Scientist* 72: 558-66.

Behrensmeyer, A. K., and A. Hill, eds. 1980. *Fossils in the Making.* Chicago: University of Chicago Press.

Behrensmeyer, A. K., and S. M. Kidwell. 1985. Taphonomy's contribution to paleobiology. *Paleobiology* 11: 105-19.

Bengtson, S. 1986. The problem of the Problematica, pp. 3-11. In *Problematic Fossil Taxa,* ed. A. Hoffman and M. H. Nitecki. New York: Oxford University Press.

Benton, M. J. 1983. Large-scale replacements in the history of life. *Nature* 302: 16-17.

Benton, M. J. 1987. Progress and competition in macroevolution. *Biological Reviews* 62: 305-38.

Bergström, J., and R. Levi-Setti. 1978. Phenotypic variation in the Middle Cambrian trilobite *Paradoxides davidis* Salter et Manuels, SE Newfoundland. *Geologica et Palaeontologica* 12: 1-40.

Bonner, J. T., ed. 1982. *Evolution and Development.* Berlin: Springer-Verlag.

Bottjer, D. J., and W. I. Ausich. 1986. Phanerozoic development of tiering in soft-substrata suspension-feeding communities. *Paleobiology* 12: 400-420.

Bottjer, D. J., and D. Jablonski. 1988. Paleoenvironmental patterns in the evolution of post-Paleozoic benthic marine invertebrates. *Palaios* 3: 540-60.

Brasier, M. D. 1979. The Cambrian radiation event, pp. 103-59. In *The Origin of Major Invertebrate Groups*, ed. M. R. House. *Systematics Association Special Volume* 12. London: Academic Press.

Brasier, M. D. 1982. Sea-level changes, facies changes and the late Precambrian-early Cambrian evolutionary explosion. *Precambrian Research* 17: 105-23.

Bretsky, P. W. 1969. Evolution of Paleozoic benthic marine invertebrate communities. *Palaeogeography, Palaeoclimatology, Palaeoecology* 6: 45-59.

Briggs, J. C. 1984. Centers of origin in biogeography. *University of Leeds, Biogeographical Monographs* 1: 95 pp.

Campbell, C. A., and J. W. Valentine. 1977. Comparability of modern and ancient marine faunal provinces. *Paleobiology* 3: 49-57.

Campbell, K. S. W., and C. R. Marshall. 1987. Rates of evolution among Palaeozoic echinoderms, pp. 61-100. In *Rates of Evolution*, ed. K. S. W. Campbell and M. F. Day. London: Allen & Unwin.

Chauff, K. M. 1985. Phylogeny of the multielement conodont genera *Bactrognathus, Doliognathus* and *Staurognathus*. *Journal of Paleontology* 59: 299-309.

Conway Morris, S. 1985. Cambrian Lagerstätten: Their distribution and significance. *Philosophical Transactions of the Royal Society of London,* B311: 49-65.

Conway Morris, S. 1989. The persistence of Burgess Shale-type faunas: Implications for the evolution of deeper-water faunas. *Transactions of the Royal Society of Edinburgh: Earth Sciences* 80: 271-83.

Conway Morris, S., and W. H. Fritz. 1984. *Lapworthella filigrana* n. sp. (incertae sedis) from the Lower Cambrian of the Cassiar Mountains, northern British Columbia, Canada, with comments on possible levels of competition in the early Cambrian. *Paläontologische Zeitschrift* 58: 197-209.

Conway Morris, S., J. S. Peel, A. K. Higgins, N. J. Soper, and N. C. Davis. 1987. A Burgess Shale-like fauna from the Lower Cambrian of north Greenland. *Nature* 326: 181-83.

Crame, J. A. 1986. Polar origins of marine invertebrate faunas. *Palaios* 1: 616-17.

Crane, P. R. 1985. Phylogenetic analysis of seed plants and the origin of angiosperms. *Annals of the Missouri Botanical Garden* 72: 716-93.

Crane, P. R. 1987. Vegetational consequences of the angiosperm diversification, pp. 107-44. In *The Origins of Angiosperms and Their Biological Consequences*, ed. E. M. Friis, W. G. Chaloner, and P. R. Crane. Cambridge: Cambridge University Press.

Crimes, T. P. 1974. Colonisation of the early ocean floor. *Nature* 248: 328-30.

Douglas, R. G., and F. Woodruff. 1981. Deep-sea benthic Foraminifera, pp. 1233-1327. In

*The Sea.* Vol. 6. *The Oceanic Lithosphere,* ed. C. Emiliani. New York: Wiley-Interscience.

Doyle, J. A. 1978. Origin of angiosperms. *Annual Review of Ecology and Systematics* 9: 365-92.

Doyle, J. A. 1984. Evolutionary, geographic, and ecological aspects of the rise of angiosperms. *Proceedings of the 27th International Geological Congress* 2: 23-33.

Doyle, J. A., and M. J. Donoghue. 1986. Seed plant phylogeny and the origin of angiosperms: An experimental cladistic approach. *Botanical Review* 52: 321-431.

Droser, M. L., and D. J. Bottjer. 1988. Trends in depth and extent of bioturbation in Cambrian carbonate marine environments, western United States. *Geology* 16: 233-36.

Eckert, J. D. 1988. Late Ordovician extinction of North American and British crinoids. *Lethaia* 21: 147-67.

Erwin, D. H., J. W. Valentine, and J. J. Sepkoski, Jr. 1987. A comparative study of diversification events: The early Paleozoic versus the Mesozoic. *Evolution* 41: 1177-86.

Fisher, D. C. 1985. Evolutionary morphology: Beyond the analogous, the anecdotal, and the ad hoc. *Paleobiology* 11: 120-38.

Fitzpatrick, J. W. 1988. Why so many passerine birds? A response to Raikow. *Systematic Zoology* 37: 71-76.

Flynn, J. J., B. J. MacFadden, and M. C. McKenna. 1984. Land-mammal ages, faunal heterochrony [*sic*], and temporal resolution in Cenozoic terrestrial sequences. *Journal of Geology* 92: 697-705.

Fortey, R. A. 1983. Cambrian-Ordovician trilobites from the boundary beds in western Newfoundland and their phylogenetic significance. In *Trilobites and Other Early Arthropods,* ed. D. E. G. Briggs and P. D. Lane. *Special Papers in Palaeontology* 30: 179-211.

Frey, R. C. 1987. The occurrence of pelecypods in Early Paleozoic epeiric-sea environments, Late Ordovician of the Cincinnati, Ohio area. *Palaios* 2: 3-23.

Gilinsky, N. L., and R. K. Bambach. 1987. Asymmetrical patterns of origination and extinction in higher taxa. *Paleobiology* 13: 427-45.

Goodwin, B. C., N. Holder, and C. C. Wylie, eds. 1983. *Development and Evolution.* Cambridge: Cambridge University Press.

Gordon, W. A. 1974. Physical controls on marine biotic distribution in the Jurassic Period. *Society of Economic Paleontologists and Mineralogists, Special Publication* 21: 136-47.

Gould, S. J. 1977. *Ontogeny and Phylogeny.* Cambridge, MA: Harvard University Press.

Gould, S. J., N. L. Gilinsky, and R. Z. German. 1987. Asymmetry of lineages and the direction of evolutionary time. *Science* 236: 1437-41.

Gould, S. J., D. M. Raup, J. J. Sepkoski, Jr., T. J. M. Schopf, and D. S. Simberloff. 1977. The shape of evolution: A comparison of real and random clades. *Paleobiology* 3: 23-40.

Hallam, A. 1987. Radiations and extinctions in relation to environmental change in the marine Lower Jurassic of northwest Europe. *Paleobiology* 13: 152-68.

Hermans, C. O. 1979. Polychaete egg sizes, life histories and phylogeny, pp. 1-9. In *Reproductive Ecology of Marine Invertebrates*, ed. S. E. Stancyk. Belle W. Baruch Library in Marine Science 9. Columbia, SC: University of South Carolina Press.

Hickey, L. J., R. M. West, M. R. Dawson, and D. K. Choi. 1983. Arctic terrestrial biota: Paleomagnetic evidence of age disparity with mid-northern latitude occurrences during the late Cretaceous and early Tertiary. *Science* 221: 1153-56.

Hoffman, A., and J. A. Kitchell. 1984. Evolution in a pelagic planktic system: A paleobiologic test of models of multispecies evolution. *Paleobiology* 10: 9-33.

Hoffman, A., and M. H. Nitecki, eds. 1986. *Problematic Fossil Taxa.* New York: Oxford University Press.

Jablonski, D. 1986. Evolutionary consequences of mass extinctions, pp. 313-29. In *Patterns and Processes in the History of Life*, ed. D. M. Raup and D. Jablonski. Berlin: Springer-Verlag.

Jablonski, D., and D. J. Bottjer. 1983. Soft-bottom epifaunal suspension-feeding communities in the Late Cretaceous: Implications for the evolution of benthic paleocommunities, pp. 747-82. In *Biotic Interactions in Recent and Fossil Benthic Communities*, ed. M. J. S. Tevesz and P. L. McCall. New York: Plenum.

Jablonski, D., and D. J. Bottjer. 1988. Onshore-offshore evolutionary patterns in post-Paleozoic echinoderms: A preliminary analysis, pp. 81-90. In *Echinoderm biology – Proceedings of the 6th International Echinoderm Conference*, ed. R. D. Burke, P. V. Mladenov, P. Lambert, and R. L. Parsley. Rotterdam: A. A. Balkema.

Jablonski, D., and D. J. Bottjer. 1990. Onshore-offshore trends in marine invertebrate evolution. In *Causes of Evolution: A Paleontological Perspective*, ed. R. M. Ross and W. D. Allmon. Chicago: The University of Chicago Press.

Jablonski, D., K. W. Flessa, and J. W. Valentine. 1985. Biogeography and paleobiology. *Paleobiology* 11: 75-90.

Jablonski, D., J. J. Sepkoski, Jr., D. J. Bottjer, and P. M. Sheehan. 1983. Onshore-offshore patterns in the evolution of Phanerozoic shelf communities. *Science* 222: 1123-25.

Jackson, J. B. C. 1974. Biogeographic consequences of eurytopy and stenotopy among marine bivalves and their evolutionary significance. *American Naturalist* 108: 541-60.

Kennett, J. P., ed. 1985. The Miocene ocean: Paleoceanography and biogeography. *Geological Society of America Memoir* 163: 337 pp.

Kidwell, S. M., and A. K. Behrensmeyer. 1988. Overview: Ecological and evolutionary implications of taphonomic processes. *Palaeogeography, Palaeoclimatology, Palaeoecology* 63: 1-13.

Kitchell, J. A., and N. MacLeod. 1988. Macroevolutionary interpretations of symmetry and synchroneity in the fossil record. *Science* 240: 1190-93.

Kitchell, J. A., and D. Pena. 1984. Periodicity of extinctions in the geologic past: Deterministic versus stochastic explanations. *Science* 226: 689-92.

Knoll, A. H. 1985. Exceptional preservation of photosynthetic organisms in silicified carbonates and silicified peats. *Philosophical Transactions of the Royal Society of London* B 311: 111-22.

Lauder, G. V. 1981. Form and function: Structural analysis in evolutionary morphology. *Paleobiology* 7: 430-42.

Levinton, J. 1988. *Genetics, Paleontology, and Macroevolution.* Cambridge: Cambridge University Press.

Liem, K. F. 1973. Evolutionary strategies and morphological innovations: Cichlid pharyngeal jaws. *Systematic Zoology* 22: 425-41.

Liem, K. F. 1980. Adaptive significance of intra- and interspecific differences in the feeding repertoires of cichlid fishes. *American Zoologist* 20: 295-314.

Lipps, J. H. 1990. Proterozoic and Cambrian Protista. In *The Proterozoic Biosphere: A Multidisciplinary Study,* ed. J. W. Schopf and C. Klein. New York: Cambridge University Press.

Lowenstam, H. A., and L. Margulis. 1980. Evolutionary prerequisites for early Phanerozoic calcareous skeletons. *BioSystems* 12: 27-41.

Macurda, D. B., Jr., and D. L. Meyer. 1983. Sea lilies and feather stars. *American Scientist* 71: 354-65.

Margalef, R. 1968. *Perspectives in Ecological Theory.* Chicago: The University of Chicago Press.

Matthews, R. K., and R. Z. Poore. 1980. Tertiary delta-[18]O record and glacioeustatic sea-level fluctuations. *Geology* 8: 501-4.

Mayr, E. 1960. The emergence of evolutionary novelties, pp. 349-80. In *Evolution after Darwin.* Vol. 1. *The Evolution of Life,* ed. S. Tax. Chicago: The University of Chicago Press.

McGhee, G. R., Jr. 1981. Evolutionary replacement of ecological equivalents in Late Devonian benthic marine communities. *Palaeogeography, Palaeoclimatology, Palaeoecology* 34: 267-83.

McKinney, M. L. 1986. Ecological causation of heterochrony: A test and implications for evolutionary theory. *Paleobiology* 12: 282-89.

McNamara, K. J. 1983. Progenesis in trilobites. In *Trilobites and Other Early Arthropods,* ed. D. E. G. Briggs and P. D. Lane. *Special Papers in Palaeontology* 30: 59-68.

McNamara, K. J. 1986a. The role of heterochrony in the evolution of Cambrian trilobites. *Biological Reviews* 61: 121-56.

McNamara, K. J. 1986b. A guide to the nomenclature of heterochrony. *Journal of Paleontology* 60: 4-13.

Meyer, D. L., and D. B. Macurda, Jr. 1977. Adaptive radiation of comatulid crinoids. *Paleobiology* 3: 74-82.

Miller, A. H. 1949. Some ecologic and morphologic considerations in the evolution of higher taxonomic categories, pp. 84-88. In *Ornithologie als biologische Wissenschaft,* ed. E. Mayr and E. Schüz. Heidelberg: Carl Winter.

Miller, A. I. 1988. Spatio-temporal transitions in Paleozoic Bivalvia: An analysis of North American fossil assemblages. *Historical Biology* 1: 251-73.

Miller, A. I., and J. J. Sepkoski, Jr. 1988. Modeling bivalve diversification: The effect of interaction on a macroevolutionary system. *Paleobiology* 14: 364-69.

Miller, K. G., R. G. Fairbanks, and G. S. Mountain. 1987. Tertiary oxygen isotope synthesis, sea level history, and continental margin erosion. *Paleoceanography* 2: 1-19.

Mount, J. F., and P. W. Signor. 1985. Early Cambrian innovation in shallow subtidal environments: Paleoenvironments of Early Cambrian shelly fossils. *Geology* 13: 730-33.

Nelson, G., and N. I. Platnick. 1981. *Systematics and Biogeography.* New York: Columbia University Press.

Palmer, A. R. 1986. Inferring relative levels of genetic variability in fossils: The link between heterozygosity and fluctuating asymmetry. *Paleobiology* 12: 1-5.

Palmer, A. R., and C. Strobeck. 1986. Fluctuating asymmetry: Measurement, analysis, patterns. *Annual Review of Ecology and Systematics* 17: 391-421.

Paul, C. R. C. 1977. Evolution of primitive echinoderms, pp. 123-58. In *Patterns of Evolution,* ed. A. Hallam. Amsterdam: Elsevier.

Paul, C. R. C. 1979. Early echinoderm radiation, pp. 415-34. In *The Origin of Major Invertebrate Groups,* ed. M.R. House. *Systematics Association Special Publication* 12. London: Academic Press.

Poore, R. Z., and R. K. Matthews. 1985. Oxygen isotope ranking of late Eocene and Oligocene planktonic foraminifers: Implications for Oligocene sea-surface temperatures and global ice volume. *Marine Micropaleontology* 9: 111-34.

Raff, R. A., and T. C. Kaufman. 1983. *Embryos, Genes, and Evolution.* New York: Macmillan.

Raff, R. A., and E. C. Raff, eds. 1987. *Development as an Evolutionary Process.* New York: Alan R. Liss.

Raikow, R. J. 1986. Why are there so many kinds of passerine birds? *Systematic Zoology* 35: 255-59.

Rasmussen, H. W. 1978. Articulata, pp. 813-998. In *Treatise on Invertebrate Paleontology. Part T, Echinodermata 2,* Vol. 3, ed. R. C. Moore and C. Teichert. Boulder, CO, and Lawrence, KS: Geological Society of America and University of Kansas Press.

Raup, D. M. 1983. On the early origins of major biologic groups. *Paleobiology* 9: 107-15.

Raup, D. M., and J. J. Sepkoski, Jr. 1986. Periodic extinction of families and genera. *Science* 231: 833-36.

Retallack, G. J., and D. L. Dilcher. 1986. Cretaceous angiosperm invasion of North America. *Cretaceous Research* 7: 227-52.

Rosen, B. R. 1984. Reef coral biogeography and climate through the Late Cainozoic: Just islands in the sun or a critical pattern of islands? pp. 201-62. In *Fossils and Climate*, ed. P. J. Brenchley. Chichester: Wiley.

Roux, M. 1987. Evolutionary ecology and biogeography of Recent stalked crinoids as a model for the fossil record, pp. 1-53. In *Echinoderm Studies 2*, ed. M. Jangoux and J. M. Lawrence. Rotterdam: A. A. Balkema.

Runnegar, B. 1982. The Cambrian explosion: Animals or fossils? *Journal of the Geological Society of Australia* 29: 395-411.

Runnegar, B. 1987. Rates and modes of evolution in the Mollusca, pp. 39-60. In *Rates of Evolution*, ed. K. S. W. Campbell and M. F. Day. London: Allen & Unwin.

Runnegar, B., and C. Bentley. 1983. Anatomy, ecology and affinities of the Australian Early Cambrian bivalve *Pojetaia runnergari* Jell. *Journal of Paleontology* 57: 73-92.

Sando, W. J. 1980. The paleoecology of Mississippian corals in the western conterminous United States. *Acta Palaeontologica Polonica* 25: 619-31.

Seilacher, A. 1956. Der Beginn des Kambriums als biologische Wende. *Neues Jahrbuch für Geologie und Paläontologie* 103: 155-80.

Sepkoski, J. J., Jr. 1979. A kinetic model of Phanerozoic taxonomic diversity. II. Early Phanerozoic families and multiple equilibria. *Paleobiology* 5: 222-51.

Sepkoski, J. J., Jr. 1981. A factor analytic description of the Phanerozoic marine fossil record. *Paleobiology* 7: 36-53.

Sepkoski, J. J., Jr. 1984. A kinetic model of Phanerozoic taxonomic diversity. III. Post-Paleozoic families and mass extinctions. *Paleobiology* 10: 246-67.

Sepkoski, J. J., Jr. 1987. Environmental trends in extinction during the Paleozoic. *Science* 235: 64-66.

Sepkoski, J. J., Jr. 1988. Alpha, beta or gamma – where does all the diversity go? *Paleobiology* 14: 221-34.

Sepkoski, J. J., Jr. 1989. Periodicity in extinction and the problem of catastrophism in the history of life. *Journal of the Geological Society, London* 146: 7-19.

Sepkoski, J. J., Jr., and A. I. Miller. 1985. Evolutionary faunas and the distribution of Paleozoic marine communities in space and time, pp. 153-90. In *Phanerozoic Diversity Patterns*, ed. J. W. Valentine. Princeton, NJ: Princeton University Press.

Sepkoski, J. J., Jr., and P. M. Sheehan. 1983. Diversification, faunal change, and community replacement during the Ordovician radiations, pp. 673-717. In *Biotic Interactions in Recent and Fossil Benthic Communities*, ed. M. J. S. Tevesz and P. L. McCall. New York: Plenum.

Sheehan, P. M. 1982. Brachiopod macroevolution at the Ordovician-Silurian boundary. *Third North American Paleontological Convention Proceedings* 2: 477-81.

Sheehan, P. M. 1986. Macroevolution and low diversity faunas. (Abstr.) *Geological Society of America Abstracts* 18: 324.

Signor, P. W. 1988. The Precambrian-Cambrian metazoan radiation: Significance of earliest Cambrian agglutinated skeletons. (Abstr.) *Geological Society of America Abstracts* 20: A104.

Simms, M. J. 1988a. The phylogeny of post-Palaeozoic crinoids, pp. 269-84. In *Echinoderm Phylogeny and Evolutionary Biology,* ed. C. R. C. Paul and A. B. Smith. Oxford: Clarendon Press.

Simms, M. J. 1988b. Patterns of evolution among Lower Jurassic crinoids. *Historical Biology* 1: 17-44.

Simpson, G. G. 1953. *The Major Features of Evolution.* New York: Columbia University Press.

Skelton, P. W. 1985. Preadaptation and evolutionary innovation in rudist bivalves. *Special Papers in Palaeontology* 33: 159-73.

Smith, A. B. 1984. Classification of the Echinodermata. *Palaeontology* 27: 431-59.

Smith, A. B. 1988. Patterns of diversification and extinction in Early Palaeozoic echinoderms. *Palaeontology* 31: 799-828.

Sohl, N. F. 1987. Cretaceous gastropods: Contrast between Tethys and the temperate provinces. *Journal of Paleontology* 61: 1085-1111.

Spicer, R. A., J. A. Wolfe, and D. J. Nichols. 1987. Alaskan Cretaceous-Tertiary floras and Arctic origins. *Paleobiology* 13: 73-83.

Sprinkle, J. 1983. Patterns and problems in echinoderm evolution, pp. 1-18. In *Echinoderm Studies 1,* ed. M. Jangoux and J. M. Lawrence. Rotterdam: A. A. Balkema.

Stanley, S. M. 1979. *Macroevolution: Pattern and Process.* San Francisco: W. H. Freeman.

Stehli, F. G. 1968. Taxonomic diversity gradients in pole location: The Recent model, pp. 163-227. In *Evolution and Environment,* ed. E. T. Drake. New Haven, CT: Yale University Press.

Stehli, F. G., R. G. Douglas, and I. A. Kagescioglu. 1972. Models for the evolution of planktonic Foraminifera, pp. 116-28. In *Models in Paleobiology,* ed. T. J. M. Schopf. San Francisco: Freeman, Cooper.

Stehli, F. G., R. G. Douglas, and N. D. Newell. 1969. Generation and maintenance of gradients in taxonomic diversity. *Science* 164: 947-49.

Stehli, F. G., A. L. McAlester, and C. E. Helsley. 1967. Taxonomic diversity of Recent bivalves and some implications for geology. *Geological Society of America Bulletin* 78: 455-66.

Stehli, F. G., and J. W. Wells. 1971. Diversity and age patterns in hermatypic corals. *Systematic Zoology* 20: 115-26.

Stenseth, N. C. 1984. The tropics: Cradle or museum? *Oikos* 43: 417-20.

Stitt, J. H. 1975. Adaptive radiation, trilobite paleoecology and extinction, Ptychaspid Biomere, Late Cambrian of Oklahoma. *Fossils and Strata* 4: 381-90.

Strathmann, R. R., and M. Slatkin. 1983. The improbability of animal phyla with few species. *Paleobiology* 9: 97-106.

Susman, R. L. 1988. Hand of *Paranthropus robustus* from Member 1, Swartkrans: Fossil evidence for tool behavior. *Science* 240: 781-84.

Sutton, R. G., and G. R. McGhee, Jr. 1985. The evolution of Frasnian marine "community-types" of south-central New York. *Geological Society of America Special Paper* 201: 211-24.

Taylor, J. D., N. J. Morris, and C. N. Taylor. 1980. Food specialization and the evolution of predatory prosobranch gastropods. *Palaeontology* 23: 375-409.

Taylor, P. D. 1988. Major radiation of cheilostome bryozoans: Triggered by the evolution of a new larval type? *Historical Biology* 1: 17-44.

Templeton, A. R. 1980. The theory of speciation via the founder principle. *Genetics* 94: 1011-38.

Templeton, A. R. 1986. The relation between speciation mechanisms and macroevolutionary pattern, pp. 497-512. In *Evolutionary Processes and Theory,* ed. S. Karlins and E. Nevo. Orlando, FL: Academic Press.

Thayer, C. W. 1973. Taxonomic and environmental stability in the Paleozoic. *Science* 182: 1242-43.

Tiffney, B. H. 1984. Seed size, dispersal syndromes, and the rise of angiosperms: Evidence and hypothesis. *Annals of the Missouri Botanical Garden* 71: 551-76.

Valentine, J. W. 1969. Patterns of taxonomic and ecological structure of the shelf benthos during Phanerozoic time. *Palaeontology* 12: 684-709.

Valentine, J. W. 1973. *Evolutionary Paleoecology of the Marine Biosphere.* Englewood Cliffs, NJ: Prentice-Hall.

Valentine, J. W. 1977. General patterns of metazoan evolution, pp. 27-57. In *Patterns of Evolution,* ed. A. Hallam. Amsterdam: Elsevier.

Valentine, J. W. 1980. Determinants of diversity in higher taxonomic categories. *Paleobiology* 6: 444-50.

Valentine, J. W. 1984. Neogene marine climate trends: Implications for biogeography and evolution of the shallow-sea biota. *Geology* 12: 647-50.

Valentine, J. W. 1986. Fossil record of the origin of Baupläne and its implications, 209-22. In *Patterns and Processes in the History of Life,* ed. D. M. Raup and D. Jablonski. Berlin: Springer-Verlag.

Valentine, J. W. 1989. The macroevolution of clade shape. In *Biotic and Abiotic Factors in Evolution,* ed. R. M. Ross and W. D. Allmon. Chicago: The University of Chicago Press.

Valentine, J. W., and D. H. Erwin. 1987. Interpreting great developmental experiments: The fossil record, pp. 71-107. In *Development as an Evolutionary Process,* ed. R. A. Raff and E. C. Raff. New York: Alan R. Liss.

Valentine, J. W., and T. D. Walker. 1986. Diversity trends within a model taxonomic hierarchy. *Physica D* 22: 31-42.

Van Valen, L. M. 1962. A study of fluctuating asymmetry. *Evolution* 16: 125-42.

Van Valen, L. M. 1971. Adaptive zones and the orders of mammals. *Evolution* 25: 420-28.

Van Valen, L. M. 1973. Are categories in different phyla comparable? *Taxon* 22: 333-73.

Van Valen, L. M. 1985. A theory of origination and extinction. *Evolutionary Theory* 7: 133-42.

Van Valen, L. M., and V. C. Maiorana. 1985. Patterns of origination. *Evolutionary Theory* 7: 107-26.

Vermeij, G. J. 1978. *Biogeography and Adaptation.* Cambridge, MA: Harvard University Press.

Vermeij, G. J. 1988. The evolutionary success of passerines: A question of semantics? *Systematic Zoology* 37: 69-71.

Voigt, E. 1985. The Bryozoa of the Cretaceous-Tertiary boundary, pp. 329-42. In *Bryozoa: Ordovician to Recent,* ed. C. Nielsen and G. P. Larwood. Fredensborg, Denmark: Olsen & Olsen.

Vrba, E. S. 1983. Macroevolutionary trends: New perspectives on the roles of adaptation and incidental effect. *Science* 221: 387-89.

Vrba, E. S., and S. J. Gould. 1986. The hierarchical expansion of sorting and selection: Sorting and selection cannot be equated. *Paleobiology* 12: 217-28.

Walker, T. D., and J. W. Valentine. 1984. Equilibrium models of evolutionary species diversity and the number of empty niches. *American Naturalist* 124: 887-99.

Wei, K.-Y., and J. P. Kennett. 1986. Taxonomic evolution of Neogene planktonic Foraminifera and paleoceanographic implications. *Paleoceanography* 1: 67-84.

Westrop, S. R., and R. Ludvigsen. 1987. Biogeographic control of trilobite mass extinction at an Upper Cambrian "biomere" boundary. *Paleobiology* 13: 84-99.

Whittington, H. B. 1980. The significance of the fauna of the Burgess Shale, Middle Cambrian, British Columbia. *Proceedings of the Geologists' Association* 91: 127-48.

Wright, S. 1949. Adaptation and selection, pp. 365-89. In *Genetics, Paleontology and Evolution,* ed. G. L. Jepsen, G. G. Simpson, and E. Mayr. Princeton, NJ: Princeton University Press.

Wright, S. 1982a. The shifting balance theory and macroevolution. *Annual Review of Genetics* 16: 1-19.

Wright, S. 1982b. Character change, speciation, and the higher taxa. *Evolution* 36: 427-43.

Zinsmeister, W. J., and R. M. Feldmann. 1984. Cenozoic high latitude heterochroneity of Southern Hemisphere marine faunas. *Science* 224: 281-83.

# INDEX